T0206233

Direct Gear Design

Direct Gear Design

Second Edition

Alexander L. Kapelevich

CRC Press
Taylor & Francis Group
Boca Raton London New York

CRC Press is an imprint of the
Taylor & Francis Group, an **Informa** business

Second edition published 2021

by CRC Press
6000 Broken Sound Parkway NW, Suite 300, Boca Raton, FL 33487-2742

and by CRC Press
2 Park Square, Milton Park, Abingdon, Oxon, OX14 4RN

© 2021 Alexander L. Kapelevich

First edition published by CRC Press 2013

CRC Press is an imprint of Taylor & Francis Group, LLC

The right of Alexander L. Kapelevich to be identified as authors
of this work has been asserted by them in accordance with sections
77 and 78 of the Copyright, Designs and Patents Act 1988.

ISBN: 978-0-367-35859-4 (hbk)
ISBN: 978-0-367-77456-1 (pbk)
ISBN: 978-1-003-17148-5 (ebk)

Typeset in Palatino
by MPS Limited, Dehradun

Dedicated to the memory of Prof. Edgar B. Vulgakov

Contents

Preface

Preface to the Second Edition

The 1st edition *Direct Gear Design* book was published in 2013. Apparently, it caught the attention of gear engineers to this innovative gear design approach, which brings noticeable improvements in custom gear drive performance by introducing nonstandard, deeply optimized gear tooth geometry solutions. Initially, Direct Gear Design® was developed for extreme gear applications – aerospace gearboxes, racing vehicle transmissions, etc. Later it became useful in many specialized gear applications like e-mobility, robotics, defense, medical devices and instruments, power generation, agriculture machines and equipment, consumer products, and many others. The most beneficial implementations of Direct Gear Design® are in gears manufactured by highly productive gear forming technologies – plastic and metal injection molding, powder metal processing, die casting, net forging, extrusion, and additive technologies. For all these gear fabrication processes, tooling cost does not depend on whether a gear tooth shape is standard or custom. As a result, gears optimized with Direct Gear Design® provide the gear drives with substantially improved performance without increasing production costs compared to the same size and material standard gears.

This 2nd edition has changed significantly. Practically every chapter of the book has been revised and updated with new results of studies of the Direct Gear Design® approach. Some less relevant parts have been removed. For example, the "Gear Design Details" chapter has been eliminated, with some of its content moved to other chapters. The "Gear Geometry Optimization" chapter includes "Tooth Macrogeometry" and "Tooth Microgeometry" sections. The "Gear Fabrication Technologies and Tooling" and "Gear Measurement" chapters of the 1st edition of the book have been combined into the "Gear Manufacturing Essentials" chapter. A new "Powder Metal and Plastic Gear Design Specifics" chapter and Epilogue are included in the 2nd edition of the book.

Chapters 5 "Gear Geometry Optimization", 6 "Stress Analysis and Rating", and Section 9.1.2.2 are written with Dr. Yuriy Shekhtman. Sections 7.2 "Power Metal Gears", 9.1.2.1 "PM Gear Processing", and 9.1.3 "Additive Gear Processing" are written by Dr. Anders Flodin.

My hope is that this book will be well understood by readers, allowing them to expand Direct Gear Design® implementations to considerably improve the operating characteristics of their gear designs.

Alexander L. Kapelevich
Shoreview, MN

Preface to the First Edition

Gears and gear drives have been known and used for millennia as critical components of mechanisms and machines. Over the last several decades the development of gearing has mostly focused in the following fields: the improvement of material, manufacturing technology and tooling, thermal treatment, tooth surface engineering and coatings, tribology and lubricants, testing technology, and diagnostics. The constant demand for high-performance gear transmissions has resulted in significant progress in gear tooth microgeometry, which defines deviation from the nominal involute surface to achieve the optimal tooth contact localization for higher load capacity and lower transmission error. However, the development of gear macrogeometry (the defining of the tooth shape and dimensional proportions) and gear design methods are traditionally based on the preselected instrumental generating rack and have remained frozen in time. The vast majority of gears are designed with the standard 20° pressure angle tooth proportions. For some demanding applications, like aerospace and automotive industries, the standard tooth proportions are altered to provide a higher transmission load capacity. Nevertheless, even for these applications, the gear design methodology has not evolved for many years.

This book introduces an alternate gear design approach called Direct Gear Design®. Developed over the past thirty years, it has been implemented in custom gear applications to maximize gear drive performance. Some segments of this book were published in technical magazines and presented at gear conferences. The successful implementations of this method, and the positive responses generated by the magazine publications and gear conference presentations, motivated me to write this book and share this knowledge and experience with the gear engineering community. In this book, the Direct Gear Design® method is presented as another engineering tool that can be beneficial for many gear drives. I tried to avoid general conclusions and recommendations, realizing that in custom gearing one solution can be beneficial for certain types of applications but could be completely unacceptable for others. For practical purposes and to facilitate the understanding of the Direct Gear Design® method for gear engineers, I used the same established standard gear nomenclature and specification as much as possible.

This book is written by an engineer, for engineers to show a beneficial alternative to the traditional way of gear design. I hope that it will expand the readers' perspective on the opportunity for further gear transmission improvements and inspire them to be open-minded in solving their practical gear design tasks.

Alexander L. Kapelevich
Shoreview, MN

Acknowledgments

This book would not have been possible were it not for the collaboration, inspiration, and support from my family, friends, colleagues. I express deep gratitude to:

- Dr. Yuriy V. Shekhtman for his constant assistance in software development and many years of friendship
- Dr. Anders Flodin (Höganäs AB) for sharing his expertise in power metal gear processing and additive technologies in this book
- Robert L. Errichello (Geartech), Yuriy Gmirya (Lockheed Martin), Hanspeter Dinner (EES KISSsoft GmbH), Prof. Karsten Stahl (FZG), Prof. Aizoh Kubo (Gear Technologies Co.), Dr. Franz J. Joachim (Joachim Gear Consulting) for support and encouragement in the writing of this book
- American Gear Manufacturer's Association for the privilege to be its member and to work in its Aerospace gearing committee
- Clients of AKGears, LLC for the opportunity to apply the Direct Gear Design® technique in their gear drive development projects
- My dear colleagues the gear experts from across the world that I had the pleasure of meeting throughout numerous international conferences; it was an honor to share and exchange ideas in the field of gear transmissions
- Nicola Sharpe and her team from CRC Press and Madhulika Jain and her team from MPS Limited for the book publication preparation

Alexander L. Kapelevich

Author Biography

Dr. Alexander L. Kapelevich is a gear design consultant at AKGears, LLC. He holds a Master's Degree in Mechanical Engineering from Moscow Aviation Institute and a Ph.D. Degree in Mechanical Engineering from Moscow State Technical University. Dr. Kapelevich has over 40 years of gear transmission research and design experience. He began his career working in the Russian Aviation Industry, where he was involved in R&D, software development, testing, and failure investigation for aerospace gear transmissions. Living in the USA since 1994, he has founded the gear design consulting firm AKGears, LLS, and developed the Direct Gear Design® methodology for custom gear transmissions, which have been implemented in various fields such as aerospace, automotive, agriculture, defense, robotics, racing, and many others. His specialty is gear drive architecture, planetary systems, gear tooth geometry optimization, gears with asymmetric teeth, and gear transmission performance maximization. Dr. Kapelevich is the author of the books titled *"Direct Gear Design"* (1st edition) and *"Asymmetric Gearing,"* as well as many technical articles. He is a member of the American Gear Manufacturers Association.

1

Historical Overview

1.1 Direct Gear Design® Origin

Gears have been known to exist and widely used for several thousand years [1]. Historically, the development of gear design dovetailed with the development and design of other mechanisms and mechanical components. Ancient engineers designed custom gears for particular applications based on the knowledge of required performance (input and output parameters) and available power sources, such as gravity, water current, wind, spring force, human or animal muscular power, etc. This knowledge allowed them to define a suitable gear arrangement and geometry, including a number of stages, location and rotation directions of input and output shafts, shape and size of the gear wheels, profile and number of teeth, and other parameters. Gear design also included material selection, which should provide the required strength and durability of every component in the gear drive.

When a gear design was completed, the next stage of gear drive development was the fabrication of parts and assembly that included technological process selection and tool design. Ancient engineers were familiar with the two most common ways to produce gears – cutting (or carving) and forming (gear forging or die casting, for example). In some cases, gear wheels and teeth (cogs or pegs) were made separately and then assembled into the body of the gear. All of these technologies defined the tool shape and process parameters using already completed gear design. Such a development sequence – design data are primary, and technology and tooling parameters are secondary – was typical for practically any mechanical component. It was also essential for gear drives. This gear design approach is called direct gear design.

Even with the earliest known use of gear mechanisms, their creators understood that gear drive performance significantly depends on gear tooth shape. The evolution of gear tooth geometry reflects a growing demand to increase load capacity, RPM, and lifetime; reduce weight and size, vibrations and noise; etc. The first primitive gears had rectangular or cylindrical teeth (Figure 1.1). These were later replaced with more advanced cycloidal tooth profiles, which are still used today in some watch and clock mechanisms.

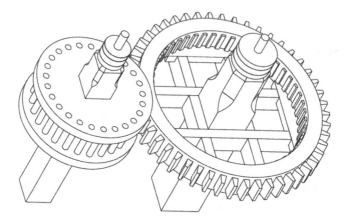

FIGURE 1.1
Ancient gear drive (Redrawn from Willis, R. 1841. *Principles of Mechanism*. London: John W. Parker, West Strand, Cambridge: J. & J.J. Deighton, p. 43).

1.2 Gear Design Based on Rack Generating Technology

In the mid-eighteenth century, Swiss scientist Leonard Euler introduced the gear tooth flank profiles formed by the involute of a circle. A critically important characteristic of the involute gear tooth profile is producing the theoretically constant rotational velocities' ratio. The involute tooth profile could be used for the external and internal gear drives as well as for the rack and pinion drives. Another vital attribute of the involute gear tooth profile is the ability of one gear to generate its mate in conjugate motion. If one of the mating gears represents a tool, it can be utilized for both the cutting (with the shaper cutter) and forming (with the gear rolling tool) fabrication processes. A gear rack is a gear wheel with an infinite number of teeth. Its tooth flank profile is a straight line. Correspondingly, a gear rack tooth flank can present a cutting edge or forming surface of the tool. The rack tool linear velocity should correspond to the rotational velocity of the mating gear blank (Figure 1.2).

Application of a tooling rack for gear manufacturing led to the invention of the gear hobbing process and machines in the nineteenth century. This invention was motivated by a massive demand for gears used in all kinds of mechanisms and machines driven by steam engines and, later, electric motors and gasoline engines. It was the time of the industrial revolution and also the beginning of industrial standardization, which considerably accelerated progress in gear development and manufacturing.

A basic gear rack is an impression of the tooling or generating rack (Figure 1.3). The main parameter of the basic gear rack is its tooth scale factor, i.e., the module m (in millimeters) in the metric system or the

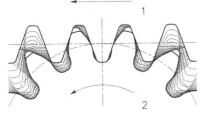

FIGURE 1.2
Rack gear generating; 1: tooling (generating) rack, 2: gear blank.

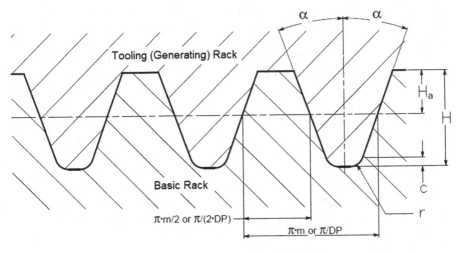

FIGURE 1.3
Basic gear rack as impression of tooling rack; α: pressure angle, H_a: addendum, H: whole depth, c: radial clearance, and r: root radius, m: module, DP: diametral pitch.

diametral pitch DP (in 1/inch) in the English system. The module m is a gear rack axial pitch divided by π or the gear pitch diameter divided by its number of teeth. The diametrical pitch DP is the number π divided by the gear rack axial pitch, or the number of gear teeth divided by its pitch diameter. The conversion formula for these scale factors is $m \times DP = 25.4$. The basic gear rack tooth flank profile angle α is called a pressure angle. For the gear, a pressure angle α is defined at the standard pitch diameter. The height of the gear tooth from the pitch diameter to the tooth tip diameter is called an addendum. An addendum divided by a module or multiplied by a diametral pitch is a dimensionless addendum coefficient. The height of the gear tooth from the pitch diameter to the tooth root diameter is called a dedendum. A dedendum divided by a module or multiplied by a diametral pitch is a dimensionless dedendum coefficient. The difference between addendum and dedendum is a radial clearance. The sum of addendum and dedendum is a whole depth. A root radius of the basic gear rack and the radial clearance divided by a module or multiplied by a diametral pitch are, consequently, the dimensionless root radius and the radial clearance coefficients. All of these gear rack tooth parameters were standardized. There is now a set of standard basic gear rack parameters that includes a pressure angle and, listed above, tooth proportion coefficients. It covers all gear tooth sizes from tiny micro gears from miniature drives to giant gears for construction and industrial machinery transmissions. Originally standard gears had a pressure angle of 14.5° because its sine is close to ¼, thus making it convenient for gear engineers to make calculations without trigonometric tables. Later, the 20° pressure angle became more common for standard gears because it provided higher load capacity and made it possible using gears with fewer teeth and without the tooth root undercut. The most common standard addendum coefficient is 1.0. The typical standard radial clearance coefficients are 0.2, 0.25, and 0.35. The standard root radius coefficients are 0.25, 0.3, and 0.38 for coarse pitch gears with the module $m \geq 1.0$ or the diametral pitch $DP \leq 20$. For the fine pitch gears ($m < 1.0$ or $DP > 20$) the root radius is not typically specified. These standard gear tooth sizes and proportion coefficients describe basic and tooling (generating) rack geometry and complete the standard gear profile for the given number of teeth.

Before the invention of the gear generating fabrication method and standardization of the basic gear rack parameters, the definition of tooling parameters was a part of the tooling design stage that followed already completed gear design. After the invention of the gear generating method, the standard tooling gear rack parameters became the input data for gear design. This made gear design indirect, depending on particular gear fabrication technology and tooling.

The standard rack generation based gear design has yielded various benefits. First of all, it is relatively simple. The given number of teeth and chosen standard tool data completely define the gearing geometry and mesh parameters. Stress analysis that results in tooth bending strength and tooth surface durability was also simplified. The predefined form and application coefficients, gear load (torque) and RPM, gear geometry parameters, and material properties define bending and contact tooth stresses. Another traditional gear design benefit is that it makes possible using a single tool for machining gears with a different number of teeth reducing tooling inventory. This gear design method also facilitates gear interchangeability for using the same standard gears for a variety of mechanisms and machines. Perhaps it was no accident that the prominent gear scientist Prof. F.L. Litvin titled one of his books *Development of Gear Technology and Theory of Gearing* [2], putting gear technology ahead of the theory of gearing. In fact, many modern gear studies are based on preselected, often standard, basic gear rack parameters. With this approach, gear design begins with the selection of the basic rack parameters. When a basic rack has selected the gear, the designer has only one parameter – the addendum modification coefficient, also known as the X-shift coefficient [3] for modification of the gear tooth geometry. The X-shift coefficient is a dimensionless factor, which is equal to the distance between the generating rack pitch line and the standard gear pitch diameter, divided by a module in the metric system or multiplied by a diametral pitch in the English system. In essence, the X-shift coefficient defines the tool position relative to the gear blank at the final cut. It indicates how far the tool is plunged into the gear blank. Figure 1.4 shows the gear tooth profiles with different values of the addendum modification coefficient.

When the X-shift coefficient is zero, the pitch line of the generating rack is tangent to the gear pitch diameter, and the gear has standard geometry. If the X-shift coefficient is less than zero (negative addendum modification), the gear's outer and root diameters, and the circular tooth thickness at the pitch diameter, are reduced. At the same time, the tooth tip land becomes larger and the load capacity of the gear with a negative addendum modification is reduced. For a low number of teeth gears, this may lead to undercutting of the involute profile at the tooth root, resulting in a contact ratio reduction and an additional reduction in tooth strength. If the X-shift coefficient is higher than zero (positive addendum modification), the gear's outer and root diameters, and the circular tooth thickness at the pitch diameter, are increased. At the same time, the tooth tip land becomes smaller. At some value of the positive X-shift coefficient for a gear with a low number of teeth, the tooth has a pointed tooth tip, which is typically unacceptable. The load capacity of the gear with a positive addendum modification is increased.

If the X-shift coefficient sum of the mating gear pair is zero, the operating pressure angle is equal to the profile angle of the tooling (generating) rack. In this case, the center distance is the same as the center distance of the standard gears with zero X-shift coefficients. This kind of an addendum modifications allows balancing the bending strength of mating gears or equalizing of the maximum specific sliding velocities of the contacting flanks, thus increasing gear mesh efficiency. A positive sum of the X-shift coefficients of mating gears increases the operating pressure angle and center distance compared to a standard gear pair. Despite some contact ratio reduction, the bending and contact (Hertzian) stresses typically become lower, resulting in increased tooth root strength and flank wear resistance. A negative sum of the X-shift coefficients of mating gears reduces the operating pressure angle and center distance compared to a standard gear pair. Despite a contact ratio increase, the

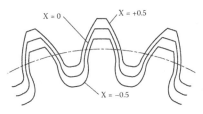

FIGURE 1.4
Gear tooth profiles with different addendum modification (X-shift) coefficients.

bending and contact (Hertzian) stresses typically become higher, resulting in reduced tooth root strength and flank wear resistance. The application of addendum modifications also allows fitting of mating gears to the given center distance.

A range of selection of the X-shift coefficients for the mating gear pair is constrained by the so-called blocking contour [4], which is created for a preselected generating rack and given numbers of teeth for mating pinion and gear. Every point of the blocking contour (Figure 1.5) with the coordinates X_1 for the pinion and X_2 for the gear defines a unique pair of mating gears. It also contains several isograms, presenting a specified tooth geometry or gear mesh conditions. The vertical line 1 and horizontal line 2 show the beginning tooth root undercut for pinion and gear, correspondingly. The isograms 3 and 4 indicate the beginning of the tip/root interference near the pinion and gear tooth root, respectively. The isogram 5 presents the gear pairs with the minimum allowable transverse contact ratio for spur gears equal to 1.0. For the practical application of the spur gears, the transverse contact ratio should be increased to 1.1, for example, as it is shown by the isogram 6. The isogram 7 represents gear pairs with a pointed tooth tip of the pinion. For practical limitation, the pinion tooth tip land must be larger than zero, e.g., $0.2 \times m$ or $0.2/DP$ (the isogram 8). The blocking contour may contain many other isograms describing different gear pair properties. For instance, the isogram 9 shows gear pairs with equalized bending stresses for both mating gears (the face widths are assumed equal). The isogram 10 presents gears with the equalized maximum

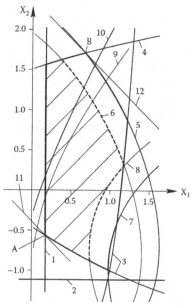

FIGURE 1.5
Blocking contour of a gear pair.

specific sliding velocities that provide maximum gear mesh efficiency. The gear pair represented by the point A of the blocking contour has the minimum operating pressure angle (the isogram 11) and the maximum contact ratio. The gear pair represented by the point B of the blocking contour has a maximum operating pressure angle (the isogram 12) and the minimum contact ratio for spur gears equal to 1.0. Some of the blocking contour border conditions, such as the tip/root interference and the pointed tooth tip isograms, cannot be violated. However, the tooth root undercut condition (if it does not reduce the transverse contact ratio below the permissible level) could be acceptable for not highly loaded gear drives. In some cases, helical gears can have a transverse contact ratio below 1.0 by compensating for it with a sufficient axial (or face) contact ratio.

Three books in the References [5–7] present a modern description of the traditional gear design approach in detail.

Traditional gear design based on the standard basic rack provides acceptable, universal solutions for many gear drive applications. However, such a universal approach does not always work satisfactorily, when maximized gear drive performance required. In this case, a gear tooth geometry optimization can provide a satisfactory solution. Search for such optimized solutions led to the development of custom basic gear racks. Some of these racks became standard for different industries. For example, the 22.5° pressure angle generating rack is frequently used to design gears for automotive transmissions. The 25° and 28° pressure angle basic racks, and the 20° pressure angle basic rack with an increased addendum (for high transverse contact ratio gears), are used to design aerospace gears [8]. Other different nonstandard basic racks are also used for gear design to meet specific application requirements.

The traditional, generating rack-based gear design has existed for more than 150 years, and its contributions to industrial progress are significant. But it becomes clear that, at its core, a rack generating design approach imposes its natural limits on gear performance improvements. Besides the generating rack machining and forming technologies, there are other highly effective gear fabrication processes, including the gear form cutting and grinding, broaching, powder metal processing, plastic and metal injection molding, precision forging and die casting, etc. All these manufacturing methods produce gears without using generating rack tools. Nevertheless, in most cases, these gears are designed utilizing the traditional generating rack-based method. And yet, despite a tremendous amount of innovation and development in gear science, technology, and machinery during the last several decades, this method is still predominantly used today by gear researchers and design engineers.

However, new gear engineering progress has diminished virtually all benefits of the traditional, indirect gear design. Technical and market performance maximization of the gear drives requires gear customization.

This modern trend does not leave so many applications for universal standard gears. Gears have become more specialized for a particular purpose and not interchangeable between different gear drives. Minimization of the tooling inventory is no longer the highest priority. Critical application gears are made using specifically dedicated cutting, forming, holding, and other tools.

Mathematical modeling, finite element analysis (FEA), and computer-aided design (CAD) software expand boundaries in the development of the optimized gear macro- and microgeometry.

1.3 Gear Design Without Rack Generation

Involute gear tooth geometry parameters can be defined without tooling rack generation. In 1934, Prof. Ch. F. Ketov described involute gear geometry parameters using a based circle diameter d_b and a number of teeth z, introducing the base module m_b [9]. Later, N.P. Lopukhov [10] further elaborated this gear parameter definition approach.

Prof. E.B. Vulgakov [11–13] originated the Gear Theory in Generalized Parameters that describes the involute gear mesh independently of gear fabrication technology and tooling generation profile. This theory is his outstanding contribution to involute gearing development. It denies the dogma about the necessity of using a preselected basic (or generating) rack profile to define involute gear parameters and gear design. Prof. E.B. Vulgakov has demonstrated how an involute gear tooth, a gear, and a gear mesh could be defined, analyzed, and optimized without using the basic or generating rack. He defined the gear mesh parameter limits and introduced the so-called areas of the existence of the gear pair. An area of existence for a particular gear pair is a significantly larger blocking contour of the same gear pair created with a generating rack. In fact, an area of existence covers gear pairs that could be generated by any possible rack (Figure 1.6), including cases when the mating gears are generated by different racks.

Indeed, the Theory of Generalized Parameters unconstrained a gear design from limitations imposed by a preselected, standard, or custom generating rack, significantly expanding the range of possible gear and mesh parameters. This theory became a foundation of modern Direct Gear Design®.

R.E. Kleiss has independently developed his own method to design of plastic gears without the use of basic or generating rack parameters [14]. He suggested using nonstandard involute tooth shapes and proportions for improving the performance of plastic gears, which is not achievable by the standard gear design, based on rack generation.

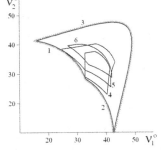

FIGURE 1.6
Area of Existence of gear pair; ν_1 and ν_2: pinion and gear involute intersection profile angles; 1: pinion root interference isogram, 2: gear root interference isogram, 3: isogram of transverse contact ratio equal 1.0; 4, 5, and 6: blocking contour borders of standard 20° and 25°, and custom 28° generating racks constructed in ν_1– –ν_2 coordinates.

1.4 Gears with Asymmetric Teeth

It is well known that the opposite gear tooth flanks are functionally different for the majority of gear drives, where the load and duration of the application are much higher for one tooth flank than for the opposite one. An asymmetric tooth shape reflects this functional difference. A design objective of asymmetric gear teeth is to improve the performance of a predominantly loaded drive tooth profile at the expense of characteristics of the opposite coast profile.

There is a common misconception about asymmetric tooth gears that they are suitable for unidirectionally rotating gear drives. In fact, asymmetric tooth gears are for unidirectionally *loaded* gear drives. There is a difference between the rotation direction and the load transmission direction. In some cases, these two directions do not coincide. For example, a gravity-loaded or spring-loaded gear drive output shaft may rotate in both directions, while the direction of loading and that for loaded gear tooth flanks remain the same. Another example: when a vehicle is moving through flat terrain or uphill, its transmission conveys power from the engine to the wheels. When it is moving downhill driven by gravity, its transmission should be shifted to the low gear to slow the vehicle down. In this case, the vehicle transmission conveys power from the wheels to the engine. In both cases, the direction of the wheels' rotation is the same regardless of terrain, but the directions of the power transmission and loaded tooth flanks are different.

Another common misconception about asymmetric unidirectionally loaded gears is that they cannot transmit load in the opposite direction by their coast tooth flanks. Besides some exceptions, which are explained in the book, the coast tooth flanks of the asymmetric unidirectionally loaded gears can transmit lesser load.

Asymmetric tooth profiles make it possible to simultaneously increase the contact ratio and operating pressure angle of the drive tooth flanks beyond those limits achievable with conventional symmetric gears. The main advantage of asymmetric tooth gears is the contact stress reduction at the drive flanks that results in greater power transmission density. Another critical advantage of the gear tooth asymmetry is the possibility of designing the coast tooth flanks independently from the drive tooth flanks, i.e., managing tooth stiffness while keeping a desirable pressure angle and contact ratio of the drive flanks. It allows increasing the tooth tip deflection, thus damping tooth mesh impact and resulting in a reduction of gear noise and vibration.

While gears with asymmetric teeth (or asymmetric gears) have been known to exist for many years, their history is sketchily described and practically unstudied. The first asymmetric teeth had a buttress shape, with the low pressure angle drive flanks and the high pressure angle coast supporting flanks to reduce tooth root stress. According to Darle W. Dudley [1]: "By 100 BC the gear art included both metal and wooden gears. Triangular teeth, buttress teeth, and pin-teeth were all in use." In about 1500 Leonardo da Vinci [15], analyzing asymmetric gear teeth (Figure 1.7), emphasized their durability: "These are a very durable type of teeth, more so than any other."

Cambridge Professor Robert Willis, in his book *Principles of Mechanism*, described the asymmetric buttress gear teeth [16]. He wrote: "If a machine be of such a nature that the wheels are only required to turn in one direction, the strength of the teeth may be doubled by an alteration of form." Prof. Willis chose the epicycloid profile for the drive tooth flanks and the involute one for the coast flanks (Figure 1.8).

In 1894, the German mechanical engineer-scientist Franz Reuleaux elaborated upon asymmetric buttress tooth shape proportions in his book *The Constructor* [17]. He also used the epicycloid drive tooth flanks and the involute coast flanks with the 53° pressure angle. He wrote: "By combining

FIGURE 1.7
Asymmetric tooth gear drawing by Leonardo da Vinci. (from Museo Galileo – Istituto e Museo di Storia Firenze Italia, with permission).

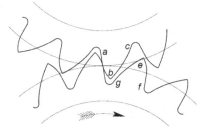

FIGURE 1.8
Asymmetric gear mesh (Redrawn from Willis, R. 1841. *Principles of Mechanism.* London: John W. Parker, West Strand, Cambridge: J. & J.J. Deighton, p. 138).

evolute and epicycloids – using the two curves for opposite sides of the same tooth – a profile of great strength is obtained. This form is of especial service for heavy-duty driving when motion is constantly in the same direction."

In 1908, Henry J. Spooner, Professor of the Polytechnic School of Engineering in London, described asymmetric teeth calling them the "Gee's buttress teeth" [18].

In 1910 Charles H. Logue, in his *American Machinist Gear Book* [19], applied involute profiles for both drive and coast flanks of the buttress asymmetric tooth gears (Figure 1.9). He wrote: "It is apparent that the object is to obtain a strong tooth or a pair of gears operating continuously in one direction ..."

In 1912 the Austrian inventor Kais Konigl patented a process for the manufacture of gear wheels with asymmetric buttress teeth [20]. The loaded tooth flank was cycloidal, and the unloaded tooth flank had an involute profile.

In 1917 Prof. O.A. Leutwiler applied involute profiles for both drive and coast flanks of the buttress or, as he called them, hook-tooth gears (Figure 1.10). He suggested the 15° pressure angle for drive flanks and the 35° pressure angle for coast flanks. He wrote [21]: "The buttress or hook-tooth gear can be used in cases where the power is always transmitted in the same

(a)   (b)

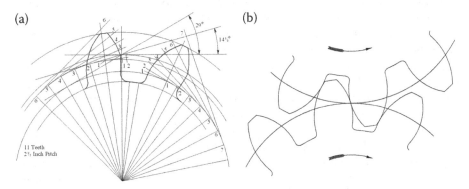

FIGURE 1.9
Buttressed tooth (a) and buttressed teeth in contact (b). (Redrawn from Logue, C.H., 1910, *American Machinist Gear Book*, McGraw-Hill Book Company, p. 30.)

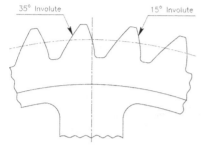

FIGURE 1.10
Buttress or "hook-tooth" gear (Redrawn from Leutwiler, O.A. 1917. *Element of Machine Design.* London: McGraw-Hill Book Company, p. 318).

direction. The load side of the tooth has the usual standard profile, while the back side of the tooth has a greater angle of obliquity."

Authors of earlier and some modern publications [23–25, for example] about asymmetric gear geometry chose the buttress tooth form that has drive flanks with low pressure angle and the coast supporting flanks with high pressure angle because it provides noticeable root bending stress reduction in comparison with the standard symmetric tooth form. However, compared to modern nonstandard optimized symmetric tooth gears, the buttress asymmetric tooth gears do not show a significant difference in root bending stress. Although, it is well known that gear transmission power density depends mainly upon the tooth flank surface durability, which is defined by the contact stress level and scuffing resistance. From this point, the application of a high pressure angle for drive tooth flanks and a low pressure angle for coast tooth flanks is more promising. In addition, this tooth form provides lower stiffness and better gear mesh impact dampening.

In most publications, gear researchers defined asymmetric gear tooth geometry traditionally generated by the preselected asymmetric gear rack (Figure 1.11). This rack is typically modified from the standard symmetric gear rack by increasing the pressure angle of one of the flanks. The opposite flank and other rack tooth proportions remain unchanged. This approach, although accustomed to standard symmetric tooth gears, and also convenient for gear machining, is not the best choice for asymmetric tooth gear design. First of all, there are no standards or standard software for gears with asymmetric teeth. Secondly, asymmetric tooth gears are for custom applications, where required performance is not achievable by using the best nonstandard optimized symmetric tooth gears. It makes essential optimizing asymmetric tooth macrogeometry (see Chapter 5), which is not possible by using an asymmetric tooling rack modified from a standard symmetric one or designed from scratch.

Prof. E.B. Vulgakov applied his Theory of Generalized Parameters to asymmetric tooth gears [13,26], defining their geometry without using rack generation parameters. According to his approach, an asymmetric tooth is constructed with two halves of the symmetric teeth with different base circles (Figure 1.12). To achieve the maximum operating pressure angle and contact ratio, the drive tooth flank is one half of the symmetric tooth, with a

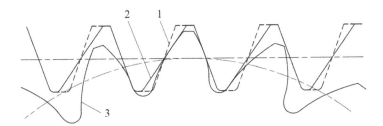

FIGURE 1.11
Asymmetric gear rack generation: 1: standard symmetric generating rack; 2: modified asymmetric generating rack profile; 3: gear profile.

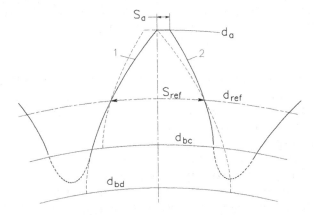

FIGURE 1.12
Asymmetric tooth constructed with two halves of the symmetric teeth; 1: drive flank from base diameter d_{bd}, 2: coast flank from base diameter d_{bc}, S_{ref}: tooth thickness at reference diameter d_{ref}, S_a: thickness at tooth tip diameter d_a, dash lines – initial symmetric tooth flanks.

pointed tip and smaller base circle. Necessary tooth tip land is provided by the coast flank that is one half of the other symmetric tooth but with a greater base circle and large tooth tip land.

The maximum transverse contact ratio and pressure angle of such asymmetric gears are the same as for symmetric gears with pointed tooth tips. This limitation does not allow for the realization of all asymmetric tooth performance improvement potentials, and this design approach did not find a practical application.

Direct Gear Design® presents an asymmetric tooth formed with two involutes of two different base circles [27–31]. Figure 1.13 shows that such an asymmetric tooth has a much longer active involute flank than the symmetric tooth with identical drive pressure angle and tooth thicknesses at the reference and tip diameters. It simultaneously enables a high drive pressure angle and required contact ratio in the asymmetric gear mesh.

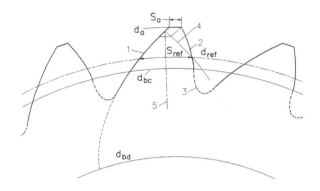

FIGURE 1.13
Asymmetric tooth constructed with two involute curves; 1: drive flank from the base diameter d_{bd}, 2: coast flank from the base diameter d_{bc}, 3: root fillet, 4: symmetric tooth profile with the same drive flank, 5: axis of the symmetric tooth, S_{ref}: tooth thickness at the reference diameter d_{ref}, S_a: tooth thickness at the tooth tip diameter d_a.

Such an approach addresses all possible asymmetric tooth profiles – from a virtually symmetric tooth shape to a tooth shape with very high asymmetry.

Charles H. Logue, describing the benefits of asymmetric tooth gears [19], wrote: "... it is a great wonder that this tooth is not more extensively used." It can be partially explained by the fact that modern gear design, which describes symmetric gear tooth macrogeometry as a result of a generating tooling rack's conjugate motion around the gear center, has been and remains the dominant gear design method for over a hundred years. Another likely reason for overlooking asymmetric gearing is the over-standardization of modern gear design. All existing gear design standards define and utilize the symmetric involute tooth profile, and the vast majority of gears have the standard 20° pressure angle and tooth proportions. However, a growing number of studies and implementations clearly indicate that asymmetric gearing has become one of the modern trends in advanced gear transmissions.

The following chapters of the book describe practically all aspects of Direct Gear Design® of symmetric and asymmetric involute gears from the tooth geometry analysis to practical applications.

2

Macrogeometry of Involute Gears

2.1 Involute and Involute Function

Involute of a circle is the path traced out by a point on the straight line that rolls around a circle. Figure 2.1 shows the involute of a circle with a radius $R = 1$. It starts at point A. A straight line normal to the involute at the point X is tangent to the circle with the center O at the point B. The angle α_x between the lines OX and OB is called a profile angle. The length of the arc AB is equal to the line BX, which is equal to $L = R \times \tan \alpha_x$. If the radius $R = 1$, $L = \tan \alpha_x$. Then the roll angle between lines OA and OB is equal to $\phi_x = L/R = \tan \alpha_x x$ (in radians). It makes the involute angle between the lines OA and OX, also called the involute function, equal to

$$inv\alpha_x = \tan \alpha_x - \alpha_x, \tag{2.1}$$

where the profile angle α_x is in radians.

The involute gear analysis utilizes the involute function and conventional trigonometric functions.

2.2 Involute Gear Tooth Parameters

2.2.1 Symmetric Gear Tooth

Two involutes of the base circle d_b unwound in opposite (clockwise and counterclockwise) directions are used to form tooth flanks (Figure 2.2). An angle α_x is the involute profile angle at some tooth flank point X. Involute function $inv\alpha_x = \tan\alpha_x - \alpha_x$ represents the angle between radial lines from the center O to the involute starting point at the circle d_b and to point X. An angle ν is the profile angle at the tooth flank intersection point.

The base tooth thickness of the external gear tooth is

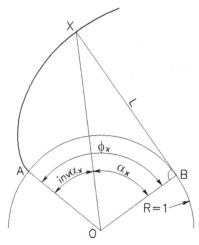

FIGURE 2.1
Involute of a circle.

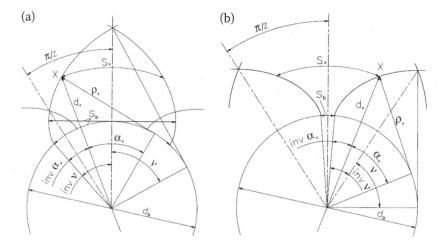

FIGURE 2.2
Involute flanks of external (a) and internal (b) gear teeth.

$$S_b = d_b \times inv(v). \qquad (2.2)$$

The base tooth thickness of the internal gear tooth is

$$S_b = d_b(\pi/z - inv(v)), \qquad (2.3)$$

where z is a number of teeth. Unlike the base tooth thickness of the external gear tooth, the base tooth thickness of the internal gear tooth depends on a number of teeth. If $inv(v) \geq \geq \pi/z$, the base tooth thickness of the internal gear tooth is equal to zero or negative.

The circular base pitch is as follows:

$$p_b = \frac{\pi d_b}{z}, \tag{2.4}$$

A diameter at the tooth flank point X is

$$d_x = d_b / \cos \alpha_x. \tag{2.5}$$

An involute profile curvature radius at the point X is

$$\rho_x = \frac{d_b \tan \alpha_x}{2}. \tag{2.6}$$

A tooth thickness at the diameter d_x is:
for external gear tooth:

$$S_x = d_x (inv(v) - inv(\alpha_x)) \tag{2.7}$$

or

$$S_x = d_b (inv(v) - inv(\alpha_x)) / \cos \alpha_x. \tag{2.8}$$

for internal gear tooth:

$$S_x = d_x (\pi/z - inv(v) + inv(\alpha_x)) \tag{2.9}$$

or

$$S_x = d_b (\pi/z - inv(v) + inv(\alpha_x)) / \cos \alpha_x. \tag{2.10}$$

A tooth profile must also include the tip land, the tooth tip radii, and the root fillet between teeth (shown with the dash lines in Figures 2.3 and 2.4). A root fillet of the tooth is not in contact with the mating gear tooth. However, it is a critically important part of the tooth profile because this is an area of the maximum root bending stress, which may limit the performance and life of a gear drive. The fillet design and optimization are presented in Section 5.1.2.

The virtual tooth tip land S_a of the external gear tooth (Figure 2.3) is defined from Equation (2.7):

$$S_a = d_a (inv(v) - inv(\alpha_a)), \tag{2.11}$$

where α_a is an involute angle at the tooth tip diameter d_a.

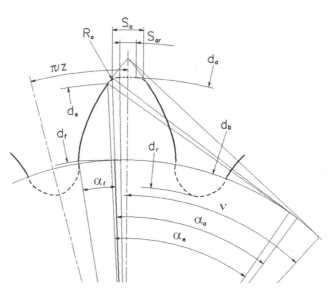

FIGURE 2.3
External gear tooth.

The virtual tooth tip land S_a of the internal gear tooth (Figure 2.4) is defined from Equation (2.9):

$$S_a = d_a(\pi/z - inv(\nu) + inv(\alpha_a)). \qquad (2.12)$$

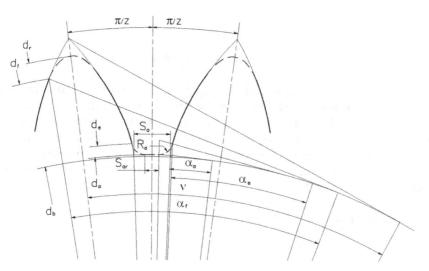

FIGURE 2.4
Internal gear tooth.

The effective involute angle at the tooth tip α_e is defined considering the tooth tip radius R_a as

$$\alpha_e = \arctan\left(\tan\left(\arccos\left(\frac{d_b}{d_a \mp 2R_a}\right)\right) \pm \frac{2R_a}{d_b}\right), \tag{2.13}$$

where the top sign (+ or –) is for the external gears and the bottom sign (+ or –) is for the internal gears.

The effective tooth tip diameter d_e (sometimes referred to as the tip form diameter) is

$$d_e = d_b / \cos \alpha_e. \tag{2.14}$$

The tooth tip land S_{ar} is defined considering the tip radius R_a:

for the external gear tooth:

$$S_{ar} = d_a (inv(\nu) - \tan \alpha_e + \arctan(\tan \alpha_e - 2R_a/d_b)), \tag{2.15}$$

for the internal gear tooth:

$$S_{ar} = d_a \left(\frac{2\pi}{z} - inv(\nu) + \tan \alpha_e - \arctan\left(\tan \alpha_e + \frac{2R_a}{d_b} \right) \right). \tag{2.16}$$

The gear form diameter d_f (sometimes referred to as the root form diameter) is at the tangent or intersection point, where the involute flank meets the root fillet. The involute angle at the gear form diameter d_f is defined as:

$$\alpha_f = \arccos(d_b / d_f). \tag{2.17}$$

The root diameter d_r is defined as a result of the fillet profile optimization (see Section 5.1.3).

A gear rack (Figure 2.5) can be considered as a gear with an infinite number of teeth. It alters an involute curve to a straight line with a constant profile angle α. The nominal pitch line equally splits the rack pitch for the tooth thickness and space between teeth. The rack nominal effective tooth addendum H_{ae} is defined from the nominal pitch line. In the rack and pinion mesh, the operating pitch line can be different from the nominal pitch line. It makes the operating pitch line tooth thickness and tooth addendum also different from the nominal ones.

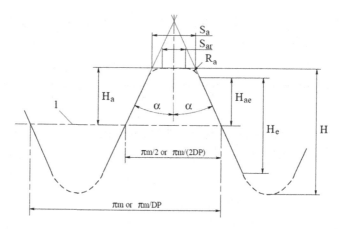

FIGURE 2.5
Gear rack tooth; 1: nominal pitch line; α: rack profile angle; m: rack module (the metric system); DP: rack diametral pitch (the English system); H_a: nominal tooth addendum; H_{ae}: nominal effective tooth addendum; H: effective tooth profile depth; H: whole tooth depth.

The tooth tip land S_{ar} of the rack tooth is:
In the metric system:

$$S_{ar} = \frac{\pi m}{2} - 2H_{ae} \tan \alpha - 2R_a \cos \alpha. \tag{2.18}$$

In the English system:

$$S_{ar} = \frac{\pi}{2DP} - 2H_{ae} \tan \alpha - 2R_a \cos \alpha. \tag{2.19}$$

The nominal gear rack tooth addendum H_a is

$$H_a = H_{ae} + R_a(1 - \sin \alpha). \tag{2.20}$$

The virtual tooth tip land S_a of the gear rack tooth is:
in the metric system:

$$S_a = \frac{\pi m}{2} - 2H_a \tan \alpha, \tag{2.21}$$

or in the English system

$$S_a = \frac{\pi}{2DP} - 2H_a \tan \alpha. \tag{2.22}$$

2.2.2 Asymmetric Gear Tooth

The application of asymmetric teeth allows improving the performance of gear drives, which transmit more load by one tooth flank in comparison to the opposite one. This type of tooth macrogeometry is practically disregarded by traditional gear design that is based on a standard gear rack with symmetric teeth. Direct Gear Design® is naturally suitable for gears with asymmetric teeth because design standards for this type of gears do not exist.

Two involute flanks of the asymmetric tooth (see Figure 2.6) are unwound from two different base diameters d_{bd} and d_{bc}. The subscripted symbol "d" is for the drive flank, and the subscripted symbol "c" is for the coast flank of an asymmetric tooth. A diameter d_x at the drive flank point X can be defined from Equation (2.5):

$$d_x = d_{bd}/\cos \alpha_{xd} = d_{bc}/\cos \alpha_{xc}, \qquad (2.23)$$

where α_{xd} and α_{xc} are the drive and coast flank involute profile angles at the diameter d_x.

Then the tooth asymmetry factor K is

$$K = d_{bc}/d_{bd} = \cos \alpha_{xc}/\cos \alpha_{xd} = \cos \nu_c/\cos \nu_d, \qquad (2.24)$$

where ν_d and ν_c are the drive and coast flank involute profile angles at the tooth flank intersection point.

For most applications of asymmetric tooth gears, the drive flank profile angle α_{xd} is greater than the coast flank profile angle α_{xc}. This means $d_{bd} < d_{bc}$ and asymmetry factor $K > 1.0$. For symmetric tooth gears $K = 1.0$.

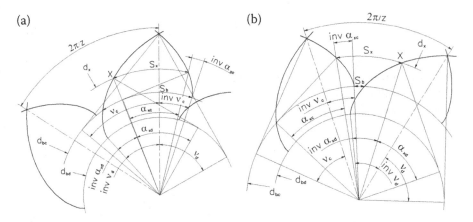

FIGURE 2.6
Involute flanks of external (a) and internal (b) asymmetric gear teeth.

At the coast flank base circle d_{bc} the coast flank profile angle $\alpha_{bc} = 0$ and the drive flank profile angle α_d from Equation (2.24) is

$$\alpha_d = \arccos(1/K). \tag{2.25}$$

The base tooth thickness of the asymmetric tooth can be defined only at the coast flank base circle d_{bc}:

for the external tooth:

$$S_b = \frac{d_{bc}}{2} \times (inv(\nu_d) + inv(\nu_c) - inv(\arccos(1/K))). \tag{2.26}$$

for the internal tooth:

$$S_b = \frac{d_{bc}}{2} \times (2\pi/z - inv(\nu_d) - inv(\nu_c) + inv(\arccos(1/K))). \tag{2.27}$$

The circular base pitches are as follows:

for the drive tooth flanks:

$$p_{bd} = \frac{\pi d_{bd}}{z}, \tag{2.28}$$

for the coast tooth flanks:

$$p_{bc} = \frac{\pi d_{bc}}{z}. \tag{2.29}$$

The relation between the drive and coast tooth flank circular base pitches is from Equation (2.24)

$$K = \frac{p_{bc}}{p_{bd}}. \tag{2.30}$$

Figure 2.7 shows the charts of the drive α_{xd} and coast α_{xc} profile angles at the different values of the asymmetry factor K.

The tooth thickness at the diameter d_x is:

for the external tooth:

$$S_x = \frac{d_x}{2} \times (inv(\nu_d) + inv(\nu_c) - inv(\alpha_{xd}) - inv(\alpha_{xc})) \tag{2.31}$$

or

$$S_x = \frac{d_{bd}}{2 \cos \alpha_{xd}} \times (inv(\nu_d) + inv(\nu_c) - inv(\alpha_{xd}) - inv(\alpha_{xc})). \tag{2.32}$$

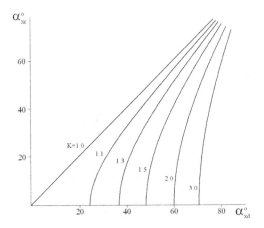

FIGURE 2.7
Relations between the drive α_{xd} and coast α_{xc} profile angles at the different values of the asymmetry factor K.

for the internal tooth:

$$S_x = \frac{d_x}{2} \times \left(\frac{2\pi}{z} - inv\,(\nu_d) - inv\,(\nu_c) + inv\,(\alpha_{xd}) + inv\,(\alpha_{xc}) \right) \qquad (2.33)$$

or

$$S_x = \frac{d_{bd}}{2\cos\alpha_{xd}} \times \left(\frac{2\pi}{z} - inv\,(\nu_d) - inv\,(\nu_c) + inv\,(\alpha_{xd}) + inv\,(\alpha_{xc}) \right). \quad (2.34)$$

The same as in a symmetric tooth, an asymmetric tooth profile must include the tip land and tip radii, and the root fillet between teeth (shown in the dash lines in the Figures 2.8–2.10).

The tooth tip diameter d_a is

$$d_a = d_{bd}/\cos\alpha_{ad} = d_{bc}/\cos\alpha_{ac}, \qquad (2.35)$$

where α_{ad} and α_{ac} are the drive and coast profile angles at the diameter d_a.

The virtual tooth tip land S_a (Figure 2.8) of the external gear tooth is defined considering tooth tip radii equal to zero from Equations (2.31) or (2.32):

$$S_a = \frac{d_a}{2} \times (inv\,(\nu_d) + inv\,(\nu_c) - inv\,(\alpha_{ad}) - inv\,(\alpha_{ac})) \qquad (2.36)$$

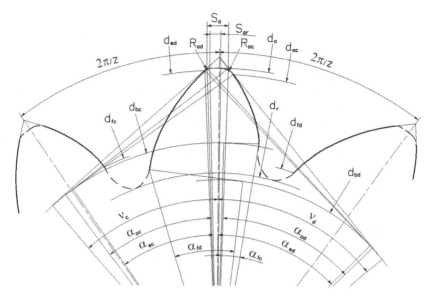

FIGURE 2.8
External asymmetric gear tooth.

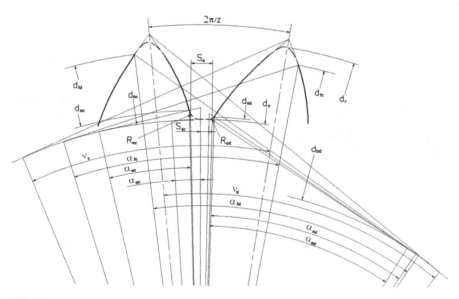

FIGURE 2.9
Internal asymmetric gear tooth.

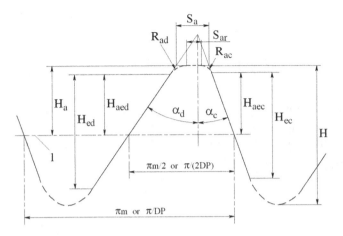

FIGURE 2.10
Asymmetric gear rack tooth; 1: nominal pitch line; α_d and α_c: rack profile angles; H_a: nominal tooth addendum; H_{aed} and H_{aec}: nominal effective tooth addendums; H_{ed} and H_{ec}: nominal effective tooth profile depths; H: whole tooth depth; m: module (in the metric System); DP: diametral pitch (in the English System).

or

$$S_a = \frac{d_{bd}}{2 \cos \alpha_{ad}} \times (inv(v_d) + inv(v_c) - inv(\alpha_{ad}) - inv(\alpha_{ac})), \qquad (2.37)$$

for internal tooth (Figure 2.9) from Equations (2.33) or (2.34):

$$S_a = \frac{d_a}{2} \times (2\pi/z - inv(v_d) - inv(v_c) + inv(\alpha_{ad}) + inv(\alpha_{ac})) \qquad (2.38)$$

or

$$S_a = \frac{d_{bd}}{2 \cos \alpha_{ad}} \times (2\pi/z - inv(v_d) - inv(v_c) + inv(\alpha_{ad}) + inv(\alpha_{ac})). \qquad (2.39)$$

The profile angles α_{ed} and α_{ec} are the effective involute angles at the tooth tip. They are defined considering the tooth tip radii R_{ad} and R_{ac} for external gears as:

for the drive flanks:

$$\alpha_{ed} = \arctan\left(\tan\left(\arccos\left(\frac{d_{bd}}{d_a \mp 2R_{ad}}\right)\right) \pm \frac{2R_{ad}}{d_{bd}}\right), \qquad (2.40)$$

for the coast flanks:

$$\alpha_{ec} = \arctan\left(\tan\left(\arccos\left(\frac{d_{bc}}{d_a \mp 2R_{ac}}\right)\right) \pm \frac{2R_{ac}}{d_{bc}}\right). \tag{2.41}$$

In Equations (2.40) and (2.41), the top sign (+ or −) is for the external gears and the bottom sign (+ or −) is for the internal gears.

The effective tooth tip diameters at the drive d_{ed} and coast d_{ec} tooth flanks are

$$d_{ed} = d_{bd}/\cos\alpha_{ed} \tag{2.42}$$

and

$$d_{ec} = d_{bc}/\cos\alpha_{ec}. \tag{2.43}$$

The tooth tip land S_{ar} of an external gear tooth is defined considering the tip radii R_{ad} and R_{ac}:

$$S_{ar} = \frac{d_a}{2} \times \left(inv(\nu_d) + inv(\nu_c) - \tan\alpha_{ed} - \tan\alpha_{ec} + \arctan\left(\tan\alpha_{ed} - \frac{2R_{ad}}{d_{bd}}\right)\right.$$

$$\left. + \arctan\left(\tan\alpha_{ec} - \frac{2R_{ac}}{d_{bc}}\right)\right) \tag{2.44}$$

The tooth tip land S_{ar} of the internal gear tooth is

$$S_{ar} = \frac{d_a}{2} \times \left(\frac{2\pi}{z} - inv(\nu_d) - inv(\nu_c) + \tan\alpha_{ed} + \tan\alpha_{ec} \right.$$

$$\left. - \arctan\left(\tan\alpha_{ed} + \frac{2R_{ad}}{d_{bd}}\right) - \arctan\left(\tan\alpha_{ec} + \frac{2R_{ac}}{d_{bc}}\right)\right). \tag{2.45}$$

The gear form diameters d_{fd} and d_{fc} at the tangent or intersection points, where the involute flanks meet the root fillet. The involute angles at the gear form diameters d_{fd} and d_{fc} are defined as:

$$\alpha_{fd} = \arccos(d_{bd}/d_{fd}) \tag{2.46}$$

and

$$\alpha_{fc} = \arccos(d_{bc}/d_{fc}). \tag{2.47}$$

The root diameter d_r is defined as a result of the fillet profile optimization (see Section 5.2).

The asymmetric gear rack tooth is shown in Figure 2.10. The tooth tip land S_{ar} of the rack tooth is:

in the metric system:

$$S_{ar} = \frac{\pi m}{2} - H_{aed} \tan \alpha_d - R_{ad} \cos \alpha_d - H_{aec} \tan \alpha_c - R_{ac} \cos \alpha_c \quad (2.48)$$

or

in the English system:

$$S_{ar} = \frac{\pi}{2DP} - H_{aed} \tan \alpha_d - R_{ad} \cos \alpha_d - H_{aec} \tan \alpha_c - R_{ac} \cos \alpha_c. \quad (2.49)$$

The nominal gear rack tooth addendum H_a is

$$H_a = H_{aed} + R_{ad}(1 - \sin \alpha_d) = H_{aec} + R_{ac}(1 - \sin \alpha_c). \quad (2.50)$$

The virtual tooth tip land S_a of the gear rack tooth is
in the metric system:

$$S_a = \frac{\pi m}{2} - H_a(\tan \alpha_d + \tan \alpha_c), \quad (2.51)$$

or

in the English system:

$$S_a = \frac{\pi}{2DP} - H_a(\tan \alpha_d + \tan \alpha_c). \quad (2.52)$$

2.3 Gear Mesh Characteristics

In this section, the gear geometry is presented, assuming that the gear tooth tip radii and the gear mesh normal backlash are equal to zero. The effects of the tooth tip radii and backlash are taken into account in the pitch factor analysis (Section 2.4) and tolerance analysis (see Section 8.3).

2.3.1 Symmetric Gearing

In traditional gear design, the module m in the metric system (or diametral pitch DP in the English system) is a scale factor defining gear tooth size. In

Direct Gear Design® the transverse operating module m_w in the metric system (or the transverse operating diametral pitch DP_w in the English system) is used as a gear tooth scale factor. It makes it understandable for engineers who are familiar with the traditional gear design. Then the nominal operating gear mesh center distance is:

$$a_w = \frac{m_w(z_2 \pm z_1)}{2} \text{ or } a_w = \frac{z_2 \pm z_1}{2DP_w}, \tag{2.53}$$

where a "+" sign is for an external gear mesh and a "−" sign is for an internal gear mesh; indexes "1" and "2" are related to parameters of the pinion with a number of teeth z_1 and of the gear with a number of teeth z_2 consequently. The pinion typically (but not always) has fewer teeth than the gear and is the driving component of the gear pair.

The operating pitch diameters of mating gears are as follows:

$$d_{w1,2} = m_w z_{1,2} \text{ or } d_{w1,2} = \frac{z_{1,2}}{DP_w}. \tag{2.54}$$

A necessary condition of a proper gear engagement is equality of the base pitches of the mating pinion and gear:

$$p_{b1} = p_{b2}. \tag{2.55}$$

2.3.1.1 Pressure Angle

Figure 2.11 shows the external, internal, and rack and pinion gear meshes. The operating circular pitch p_w is

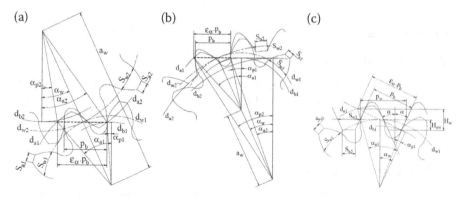

FIGURE 2.11
Symmetric gear mesh; (a) external, (b) internal, (c) rack and pinion.

$$p_w = \frac{\pi \times d_{w1,2}}{z_{1,2}} = \frac{\pi \times d_{b1,2}}{z_{1,2} \times \cos \alpha_w}, \tag{2.56}$$

where $d_{w1,2}$ – operating pitch diameters of the pinion and gear; $d_{b1,2}$ – base diameters of the pinion and gear, α_w – operating pressure angle.

It can be presented in the metric system:

$$p_w = \pi m_w, \tag{2.57}$$

or in the English system:

$$p_w = \frac{\pi}{DP_w}. \tag{2.58}$$

It is also equal to

$$p_w = S_{w1} + S_{w2}, \tag{2.59}$$

where S_{w1} and S_{w2} are the pinion and gear tooth thicknesses at the nominal operating pitch diameters d_{w1} and d_{w2} consequently.

The pinion tooth thickness S_{w1} is defined from Equation (2.8) as

$$S_{w1} = \frac{d_{b1}}{\cos \alpha_w} (inv(v_1) - inv(\alpha_w)). \tag{2.60}$$

The mating gear tooth thickness S_{w2} is:

for the external gear mesh:

$$S_{w2} = \frac{d_{b2}}{\cos \alpha_w} (inv(v_2) - inv(\alpha_w)), \tag{2.61}$$

or

for the internal gear mesh from (2.10):

$$S_{w2} = \frac{d_{b2}}{\cos \alpha_w} \left(\frac{\pi}{z_2} - inv(v_2) + inv(\alpha_w) \right). \tag{2.62}$$

For the rack and pinion mesh, the operating module m_w is equal to the rack module m in the metric system, or the operating diametral pitch DP_w is equal to the rack diametral pitch DP in the English system. The mating gear rack tooth thickness at the operating pitch line S_{w2} is

$$S_{w2} = \pi m - S_{w1} \text{ or } S_{w2} = \frac{\pi}{DP} - S_{w1}. \tag{2.63}$$

Considering Equation (2.60), it is:
   in the metric system:

$$S_{w2} = \pi m - \frac{d_{b1}}{\cos \alpha_w}(inv(\nu_1) - inv(\alpha_w)), \qquad (2.64)$$

or
   in the English system:

$$S_{w2} = \frac{\pi}{DP} - \frac{d_{b1}}{\cos \alpha_w}(inv(\nu_1) - inv(\alpha_w)). \qquad (2.65)$$

The rack tooth operating addendum is

$$H_{aw} = \frac{S_{w2} - S_{a2}}{2 \tan \alpha}. \qquad (2.66)$$

The nominal operating pressure angle α_w is defined [11] by substitution of S_{w1} and S_{w2} from Equations (2.60) and (2.61) or (2.62) into (2.59) considering also (2.56):
   for the external gear mesh:

$$inv(\alpha_w) = \frac{1}{1 + u}\left(inv(\nu_1) + u \times inv(\nu_2) - \frac{\pi}{z_1}\right), \qquad (2.67)$$

or
   for the internal gear mesh:

$$inv(\alpha_w) = \frac{1}{u - 1}(u \times inv(\nu_2) - inv(\nu_1)), \qquad (2.68)$$

where $u = z_2/z_1$, the gear ratio.
   In the rack and pinion mesh the nominal operating pressure angle α_w is equal to the rack profile angle α.

2.3.1.2 Tip/Root Interference

The profile angles at the tooth contact points near the root fillets (see Figure 2.11) are [11]:
   for the external gear mesh:

$$\alpha_{p1} = \arctan((1 + u)\tan \alpha_w - u \tan \alpha_{a2}), \qquad (2.69)$$

$$\alpha_{p2} = \arctan\left(\frac{1+u}{u}\tan\alpha_w - \frac{1}{u}\tan\alpha_{a1}\right), \tag{2.70}$$

for the internal gear mesh:

$$\alpha_{p1} = \arctan(u\tan\alpha_{a2} - (u-1)\tan\alpha_w), \tag{2.71}$$

$$\alpha_{p2} = \arctan\left(\frac{u-1}{u}\tan\alpha_w + \frac{1}{u}\tan\alpha_{a1}\right), \tag{2.72}$$

for the rack and pinion mesh:

$$\alpha_{p1} = \arctan\left(\tan\alpha - \frac{2H_{aw}}{d_{b1}\sin\alpha}\right), \tag{2.73}$$

$$\alpha_{p2} = \alpha. \tag{2.74}$$

The rack tooth operating depth is

$$H_w = \frac{d_{b1}\sin\alpha}{2}\left(\tan\alpha_{a1} - \tan\alpha_{p1}\right). \tag{2.75}$$

If the profile angle α_{p1} or α_{p2} in the external mesh or angle α_{p1} in the internal and rack and pinion meshes is less than zero, it is an indication of tip/root interference. It means the mating gear tooth tip undercuts the involute flank near the base circle. This type of undercut is different than in the traditional gear design where the involute tooth flank near to the tooth root is undercut by the generating (tooling) rack tooth tip.

In an external gear mesh, the pinion profile angle α_{u1} at the undercut point is defined (Figure 2.12) by the equation system

$$\begin{aligned}
\frac{\sin(inv(\alpha_w) - inv(\alpha_{u1}))}{\cos\alpha_{u1}} &= (1+u)\frac{\sin\phi_1}{\cos\alpha_w} \\
&\quad - \frac{u}{\cos\alpha_{a2}}\sin\left(\phi_1\left(1+\frac{1}{u}\right) - inv(\alpha_{a2}) + inv(\alpha_w)\right) \\
\frac{\cos(inv(\alpha_w) - inv(\alpha_{u1}))}{\cos\alpha_{u1}} &= (1+u)\frac{\cos\phi_1}{\cos\alpha_w} \\
&\quad - \frac{u}{\cos\alpha_{a2}}\cos\left(\phi_1\left(1+\frac{1}{u}\right) - inv(\alpha_{a2}) + inv(\alpha_w)\right).
\end{aligned} \tag{2.76}$$

Similarly, in the case of the gear profile undercut, the angle α_{u2} is defined by the system

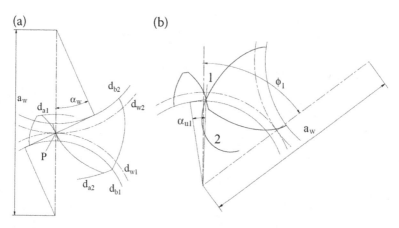

FIGURE 2.12
Definition of the undercut profile angle α_{u1} in external mesh; (a) initial tooth mesh position – gear teeth in contact at the pitch point P; (b) undercut position; 1: undercut profile point, 2: trajectory of the mating gear tooth tip.

$$\frac{u\sin(inv(\alpha_w) - inv(\alpha_{u2}))}{\cos\alpha_{u2}} = (1 + u)\frac{\sin\phi_2}{\cos\alpha_w} - \frac{\sin(\phi_2(1 + u) - inv(\alpha_{a1}) + inv(\alpha_w))}{\cos\alpha_{a1}}$$

$$\frac{u\cos(inv(\alpha_w) - inv(\alpha_{u2}))}{\cos\alpha_{u2}} = (1 + u)\frac{\cos\phi_2}{\cos\alpha_w} - \frac{\cos(\phi_2(1 + u) - inv(\alpha_{a1}) + inv(\alpha_w))}{\cos\alpha_{a1}} \qquad (2.77)$$

In an internal mesh the pinion profile angle α_{u1} at the undercut point is defined (Figure 2.13) by the system

$$\frac{\sin(inv(\alpha_w) - inv(\alpha_{u1}))}{\cos\alpha_{u1}} = \frac{u}{\cos\alpha_{a2}}\sin\left(\phi_1\left(1 - \frac{1}{u}\right) - inv(\alpha_{a2}) + inv(\alpha_w)\right)$$

$$- (u - 1)\frac{\sin\phi_1}{\cos\alpha_w},$$

$$\frac{\cos(inv(\alpha_w) - inv(\alpha_{u1}))}{\cos\alpha_{u1}} = \frac{u}{\cos\alpha_{a2}}\cos\left(\phi_1\left(1 - \frac{1}{u}\right) - inv(\alpha_{a2}) + inv(\alpha_w)\right) \qquad (2.78)$$

$$- (u - 1)\frac{\cos\phi_1}{\cos\alpha_w}.$$

For the gear with internal teeth, the angle α_{p2} is always greater than zero, and this kind of interference is impossible.

For a rack and pinion mesh, the pinion profile angle α_{u1} at the undercut point is defined (Figure 2.14) by the equation system

$$\frac{d_{b1}\phi_1}{2\cos\alpha} - H_a\tan\alpha = \frac{d_{b1}}{2\cos\alpha_{u1}}\sin(\phi_1 - inv(\alpha) - inv(\alpha_{u1})),$$

$$\frac{d_{b1}}{2\cos\alpha} - H_a = \frac{d_{b1}}{2\cos\alpha_{u1}}\cos(\phi_1 - inv(\alpha) - inv(\alpha_{u1})). \qquad (2.79)$$

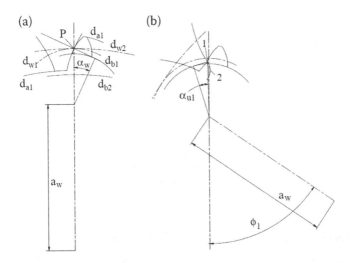

FIGURE 2.13
Definition of the undercut profile angle α_{u1} in internal mesh; (a) initial tooth mesh position – gear teeth in contact at the pitch point P; (b) undercut position; 1: undercut profile point, 2: trajectory of the mating gear tooth tip.

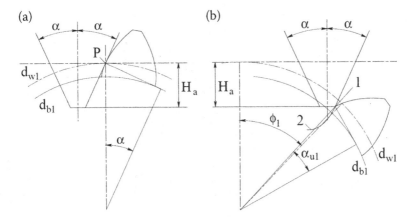

FIGURE 2.14
Definition of the undercut profile angle α_{u1} in rack and pinion mesh; (a) initial tooth mesh position – rack and pinion teeth in contact at the pitch point P; (b) undercut position; 1: undercut profile point, 2: trajectory the mating rack tooth tip.

The rack teeth do not have this kind of interference with undercut near the root fillet because $\alpha_{p2} = \alpha \geq 0$.

If the tip/root interference occurs, the lowest contact point coincides with the undercut point, and the profile angles $\alpha_{p1,2}$ become equal to profile angles at the undercut point $\alpha_{u1,2}$.

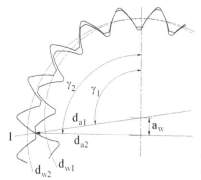

FIGURE 2.15
Tip/tip interference in the internal mesh; 1: the interference point.

2.3.1.3 Tip/Tip Interference in Internal Gearing

There is another kind of interference that can occur in internal gearing. At some gear geometry parameter combinations, tips of the mating gears may interfere (Figure 2.15).

The interference absence condition is

$$\Delta = \lambda_1 - u\lambda_2 \geq 0 \tag{2.80}$$

where

$$\lambda_{1,2} = \gamma_{1,2} + inv\,(\alpha_{a1,2}) - inv\,(\alpha_w), \tag{2.81}$$

$$\gamma_1 = \pi - \arccos\left(\frac{\frac{d_{a1}^2}{4} + a_w{}^2 - \frac{d_{a2}^2}{4}}{d_{a1}a_w}\right), \tag{2.82}$$

$$\gamma_2 = \arccos\left(\frac{\frac{d_{a2}^2}{4} + a_w{}^2 - \frac{d_{a1}^2}{4}}{d_{a2}a_w}\right). \tag{2.83}$$

This kind of interference is more typical for internal gearing with a low tooth number difference $z_2 - z_1$ and low operating pressure angle α_w. It can be avoided by increasing the pressure angle and reducing the tooth height.

2.3.1.4 Transverse Contact Ratio

The transverse contact ratio is defined as the contact line length (the dashed line in Figure 2.11) divided by the base pitch p_b, that is from Equation (2.4),

$$p_b = \frac{\pi d_{b1}}{z_1} = \frac{\pi d_{b2}}{z_2} = p_w \cos \alpha_w. \tag{2.84}$$

Then the transverse contact ratio is [11]:
for the external gear mesh:

$$\varepsilon_\alpha = \frac{z_1}{2\pi}(\tan \alpha_{a1} + u \tan \alpha_{a2} - (1 + u)\tan \alpha_w), \tag{2.85}$$

for the internal gear mesh:

$$\varepsilon_\alpha = \frac{z_1}{2\pi}(\tan \alpha_{a1} - u \tan \alpha_{a2} + (u - 1)\tan \alpha_w), \tag{2.86}$$

for the rack and pinion mesh:

$$\varepsilon_\alpha = \frac{z_1}{2\pi}\left(\tan \alpha_{a1} - \tan \alpha + \frac{2H_a}{d_{b1} \sin \alpha}\right)$$

or

$$\varepsilon_\alpha = \frac{z_1}{2\pi}(\tan \alpha_{a1} - \tan \alpha) + \frac{H_a}{p_w \tan \alpha}. \tag{2.87}$$

Alternatively, the contact ratio can be defined:
for the external gear mesh with Equations (2.69) and (2.70):

$$\varepsilon_\alpha = \frac{z_1}{2\pi}\left((1 + u)\tan \alpha_w - \tan \alpha_{p1} - u \tan \alpha_{p2}\right), \tag{2.88}$$

for the internal gear mesh with Equations (2.71) and (2.72):

$$\varepsilon_\alpha = \frac{z_1}{2\pi}\left(u \tan \alpha_{p2} - \tan \alpha_{p1} - (u - 1)\tan \alpha_w\right), \tag{2.89}$$

for the rack and pinion mesh with Equations (2.73) and (2.74):

$$\varepsilon_\alpha = \frac{z_1}{2\pi}\left(\tan \alpha - \tan \alpha_{p1} + \frac{2(H_w - H_{aw})}{d_{b1} \sin \alpha}\right), \tag{2.90}$$

where H_w is the effective rack tooth depth.

The solution of Equations (2.69), (2.70), and (2.85) for the external gears, and Equations (2.71), (2.72), and (2.86) for the internal gears defines the transverse contact ratio as follows:

$$\varepsilon_\alpha = \frac{z_1}{2\pi}(\tan \alpha_{a1} - \tan \alpha_{p1}) \tag{2.91}$$

or

$$\varepsilon_\alpha = \frac{z_2}{2\pi}(\pm \tan \alpha_{a2} \mp \tan \alpha_{p2}), \tag{2.92}$$

where the top sign (+ or −) is for the external gears and the bottom sign (+ or −) is for the internal gears.

For a rack and pinion mesh the transverse contact ratio can be also defined by solving Equations (2.73), (2.74), (2.87), and (2.90):

$$\varepsilon_\alpha = \frac{H_w}{p_b \sin \alpha} = \frac{H_w}{p_w \tan \alpha}. \tag{2.93}$$

Then for the rack and pinion mesh, the transverse contact ratio is:
for the metric system:

$$\varepsilon_\alpha = \frac{H_w}{\pi m_w \tan \alpha}. \tag{2.94}$$

for the English system:

$$\varepsilon_\alpha = \frac{DP \times H_w}{\pi \tan \alpha}. \tag{2.95}$$

2.3.2 Asymmetric Gearing

In all figures and equations describing gears with asymmetric teeth, the subscripted symbol "*d*" *is* for parameters related to the drive tooth flank and the subscripted symbol "*c*" *is* for parameters related to the coast tooth flank.

A necessary condition of a proper asymmetric gear engagement is equality of the drive and coast flank base pitches of the mating pinion and gear:

$$p_{bd1} = p_{bd2} \text{ and } p_{bc1} = p_{bc2}. \tag{2.96}$$

2.3.2.1 Pressure Angles

Figure 2.16 illustrates the definition of the drive α_{wd} and coast α_{wc} pressure angles in the asymmetric external, internal, and rack and pinion gear meshes.

The pinion and gear tooth thicknesses S_{w1} and S_{w2} at the operating pitch diameters $d_{w1,2}$ are defined by Equations (2.32) and (2.34) as follows:

$$S_{w1} = \frac{d_{bd1}}{2 \cos \alpha_{wd}} (inv(\nu_{d1}) + inv(\nu_{c1}) - inv(\alpha_{wd}) - inv(\alpha_{wc})), \quad (2.97)$$

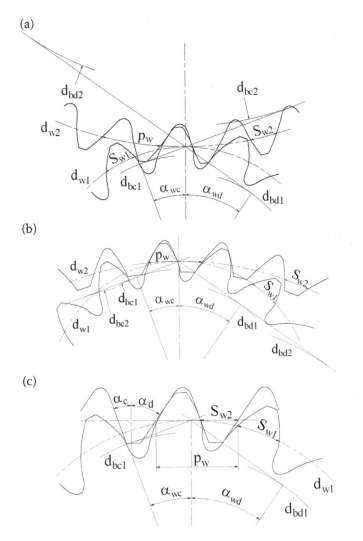

FIGURE 2.16
Pressure angles in the asymmetric gear mesh: (a) external, (b) internal, (c) rack and pinion.

for the external gearing:

$$S_{w2} = \frac{d_{bd2}}{2 \cos \alpha_{wd}} (inv(\nu_{d2}) + inv(\nu_{c2}) - inv(\alpha_{wd}) - inv(\alpha_{wc})), \quad (2.98)$$

for the internal gearing:

$$S_{w2} = \frac{d_{bd2}}{2 \cos \alpha_{wd}} \left(\frac{2\pi}{z_2} - inv(\nu_{d2}) - inv(\nu_{c2}) + inv(\alpha_{wd}) + inv(\alpha_{wc}) \right), (2.99)$$

where $d_{b1,2}$ and $d_{c1,2}$ – drive and coast base diameters of the pinion and gear, $\nu_{d1,2}$ and $\nu_{c1,2}$ – drive and coast profile angles of the pinion and gear at the tooth flank intersection points; α_{wd} and α_{wc} – drive and coast operating pressure angles.

The gear rack tooth thickness at the operating pitch line S_{w2} is:

for the metric system:

$$S_{w2} = \pi m - \frac{d_{bd1}}{2 \cos \alpha_d} (inv(\nu_{d1}) + inv(\nu_{c1}) - inv(\alpha_d) - inv(\alpha_c)), \quad (2.100)$$

for the English system:

$$S_{w2} = \frac{\pi}{DP} - \frac{d_{bd1}}{2 \cos \alpha_d} (inv(\nu_{d1}) + inv(\nu_{c1}) - inv(\alpha_d) - inv(\alpha_c)). \quad (2.101)$$

The drive flanks α_{wd} and the coast flanks α_{wc} operating pressure angles are defined by substitution of S_{w1} and S_{w2} from Equations (2.97) and (2.98) or (2.99) into (2.59):

for the external gearing:

$$inv(\alpha_{wd}) + inv(\alpha_{wc}) = \frac{1}{1 + u} \left(inv(\nu_{d1}) + inv(\nu_{c1}) + u(inv(\nu_{d2}) + inv(\nu_{c2})) - \frac{2\pi}{z_1} \right), \quad (2.102)$$

for the internal gearing:

$$inv(\alpha_{wd}) + inv(\alpha_{wc}) = \frac{1}{u - 1} (u(inv(\nu_{d2}) + inv(\nu_{c2})) - inv(\nu_{d1}) - inv(\nu_{c1})). \quad (2.103)$$

The relation between pressure angles for the drive flanks α_{wd} and pressure angles for the coast flanks α_{wc} is defined from Equation (2.24) as follows:

$$\cos \alpha_{wc} = K \times \cos \alpha_{wd}. \quad (2.104)$$

In the rack and pinion mesh the operating pressure angles α_{wd} and α_{wc} are equal to the rack profile angles α_d and α_c:

$$\alpha_{wd} = \alpha_d \text{ and } \alpha_{wc} = \alpha_c. \tag{2.105}$$

2.3.2.2 Interference of Asymmetric Gears

Figure 2.17 shows definition of the profile angles $\alpha_{pd1,2}$ and $\alpha_{pc1,2}$ at the points of tooth contact, near the root fillet in the asymmetric external, internal, and rack and pinion meshes.

The profile angles at the tooth contact points near the root fillets (Figure 2.17) are as follows:

for the external gear mesh, drive flanks:

$$\alpha_{pd1} = \arctan((1 + u)\tan\alpha_{wd} - u\tan\alpha_{ad2}), \tag{2.106}$$

$$\alpha_{pd2} = \arctan\left(\frac{1+u}{u}\tan\alpha_{wd} - \frac{1}{u}\tan\alpha_{ad1}\right), \tag{2.107}$$

for the external gear mesh, coast flanks:

$$\alpha_{pc1} = \arctan((1 + u)\tan\alpha_{wc} - u\tan\alpha_{ac2}), \tag{2.108}$$

$$\alpha_{pc2} = \arctan\left(\frac{1+u}{u}\tan\alpha_{wc} - \frac{1}{u}\tan\alpha_{ac1}\right), \tag{2.109}$$

for the internal gear mesh, drive flanks:

$$\alpha_{pd1} = \arctan(u\tan\alpha_{ad2} - (u - 1)\tan\alpha_{wd}), \tag{2.110}$$

$$\alpha_{pd2} = \arctan\left(\frac{u-1}{u}\tan\alpha_{wd} + \frac{1}{u}\tan\alpha_{ad1}\right), \tag{2.111}$$

for the internal gear mesh, coast flanks:

$$\alpha_{pc1} = \arctan(u\tan\alpha_{ac2} - (u - 1)\tan\alpha_{wc}), \tag{2.112}$$

$$\alpha_{pc2} = \arctan\left(\frac{u-1}{u}\tan\alpha_{wc} + \frac{1}{u}\tan\alpha_{ac1}\right), \tag{2.113}$$

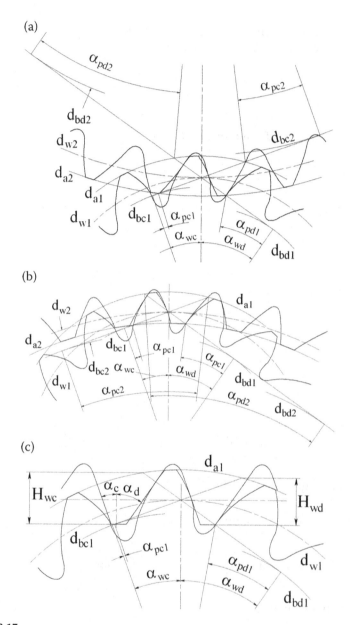

FIGURE 2.17

Profile angles $\alpha_{pd1,2}$ and $\alpha_{pc1,2}$ at the points of the tooth contact, near the root fillet in an asymmetric gear mesh; (a) external, (b) internal, (c) rack and pinion.

for the rack and pinion mesh, drive flanks:

$$\alpha_{pd1} = \arctan\left(\tan \alpha_d - \frac{2H_a}{d_{bd1} \sin \alpha_d}\right), \tag{2.114}$$

$$\alpha_{pd2} = \alpha_d, \tag{2.115}$$

for the rack and pinion mesh, coast flanks:

$$\alpha_{pc1} = \arctan\left(\tan \alpha_c - \frac{2H_a}{d_{bc1} \sin \alpha_c}\right), \tag{2.116}$$

$$\alpha_{pc2} = \alpha_c. \tag{2.117}$$

The rack tooth operating depths are

$$H_{wd} = \frac{d_{bd1} \sin \alpha_d}{2}(\tan \alpha_{ad1} - \tan \alpha_{pd1}), \tag{2.118}$$

$$H_{wc} = \frac{d_{bc1} \sin \alpha_c}{2}(\tan \alpha_{ac1} - \tan \alpha_{pc1}). \tag{2.119}$$

Since the asymmetry factor K is usually larger than 1.0, interference occurs first at the coast involute flanks. If the coast profile angle α_{pc1} or α_{pc2} in the external gear mesh, or coast profile angle α_{pc1} in the internal and rack and pinion meshes is less than zero, it is an indication of the tip/root interference. It leads to the coast profile undercut near the coast base circle.

In the external asymmetric gear mesh the pinion profile angle α_{uc1} at the undercut point is defined (Figure 2.18) by the equation system

$$\frac{\sin(inv(\alpha_{wc}) - inv(\alpha_{uc1}))}{\cos \alpha_{uc1}} = (1 + u)\frac{\sin \phi_1}{\cos \alpha_{wc}}$$

$$- \frac{u}{\cos \alpha_{ac2}}\sin\left(\phi_1\left(1 + \frac{1}{u}\right) - inv(\alpha_{ac2}) + inv(\alpha_{wc})\right),$$

$$\frac{\cos(inv(\alpha_{wc}) - inv(\alpha_{u1}))}{\cos \alpha_{uc1}} = (1 + u)\frac{\cos \phi_1}{\cos \alpha_{wc}}$$

$$- \frac{u}{\cos \alpha_{ac2}}\cos\left(\phi_1\left(1 + \frac{1}{u}\right) - inv(\alpha_{ac2}) + inv(\alpha_{wc})\right). \tag{2.120}$$

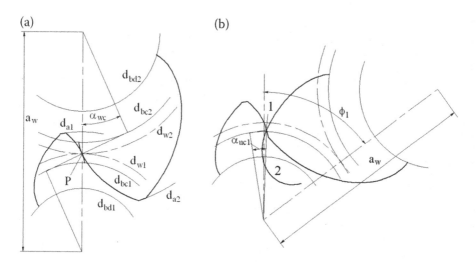

FIGURE 2.18
Definition of the undercut profile angle α_{uc1} in external mesh; (a) initial tooth mesh position – gear teeth in contact at the pitch point P; (b) undercut position; 1: undercut profile point, 2: trajectory of the mating gear tooth tip.

Similar, in case of the gear profile undercut, the angle α_{uc2} is defined by the system

$$
\begin{aligned}
\frac{u\sin(inv(\alpha_w) - inv(\alpha_{uc2}))}{\cos\alpha_{uc2}} &= (1+u)\frac{\sin\phi_2}{\cos\alpha_{wc}} - \frac{\sin(\phi_2(1+u) - inv(\alpha_{ac1}) + inv(\alpha_{wc}))}{\cos\alpha_{ac1}}, \\
\frac{u\cos(inv(\alpha_w) - inv(\alpha_{u2}))}{\cos\alpha_{uc2}} &= (1+u)\frac{\cos\phi_2}{\cos\alpha_{wc}} - \frac{\cos(\phi_2(1+u) - inv(\alpha_{ac1}) + inv(\alpha_{wc}))}{\cos\alpha_{ac1}}.
\end{aligned}
\tag{2.121}
$$

In the internal asymmetric gear mesh the pinion profile angle α_{uc1} at the undercut point is defined (Figure 2.19) by the system

$$
\begin{aligned}
\frac{\sin(inv(\alpha_{wc}) - inv(\alpha_{uc1}))}{\cos\alpha_{uc1}} &= \frac{u}{\cos\alpha_{ac2}}\sin\left(\phi_1\left(1-\frac{1}{u}\right) - inv(\alpha_{ac2}) + inv(\alpha_{wc})\right) \\
&\quad - (u-1)\frac{\sin\phi_1}{\cos\alpha_{wc}}, \\
\frac{\cos(inv(\alpha_{wc}) - inv(\alpha_{uc1}))}{\cos\alpha_{uc1}} &= \frac{u}{\cos\alpha_{ac2}}\cos\left(\phi_1\left(1-\frac{1}{u}\right) - inv(\alpha_{ac2}) + inv(\alpha_{wc})\right) \\
&\quad - (u-1)\frac{\cos\phi_1}{\cos\alpha_{wc}}.
\end{aligned}
\tag{2.122}
$$

For the internal tooth gear the angle α_{pc2} is always greater than zero and this kind of interference with undercut near the gear tooth fillet is not possible.

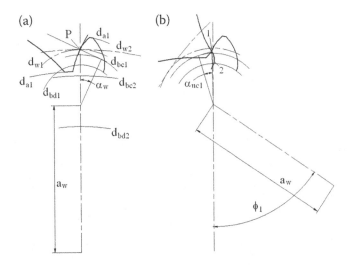

FIGURE 2.19
Definition of the undercut profile angle α_{uc1} in internal mesh; (a) initial tooth mesh position – gear teeth in contact at the pitch point P; (b) undercut position; 1: undercut profile point, 2: trajectory of the mating gear tooth tip.

For the rack and pinion mesh the pinion profile angle α_{uc1} at the undercut point is defined (Figure 2.20) by the equation system

$$\frac{d_{bc1}\phi_1}{2\cos\alpha} - H_a\tan\alpha_c = \frac{d_{bc1}}{2\cos\alpha_{uc1}}\sin(\phi_1 - inv(\alpha_c) - inv(\alpha_{uc1}))$$
$$\frac{d_{bc1}}{2\cos\alpha_c} - H_a = \frac{d_{bc1}}{2\cos\alpha_{uc1}}\cos(\phi_1 - inv(\alpha_c) - inv(\alpha_{uc1}))$$

(2.123)

The rack teeth do not have this kind of interference with undercut near the fillet, because $\alpha_{p2} = \alpha \geq 0$.

If an undercut occurs, the lowest contact point coincides with the undercut point, and the coast profile angles $\alpha_{pc1,2}$ become equal to the coast profile angles at the undercut point $\alpha_{uc1,2}$.

The tip/tip interference is typical for the internal gearing with a low tooth number difference $z_2 - z_1$. At some gear geometry parameter combination, the tips of the mating gears may interfere (Figure 2.21). In the asymmetric tooth gear mesh, such a tip/tip interference occurs first at the low pressure angle flanks. It can be avoided by increasing the coast flank pressure angle or reducing the tooth height .

The interference absence condition is described by Equations (2.80–2.83) for the symmetric internal gearing. But they are also applicable to the low pressure angle flanks of the internal asymmetric gears to define their interference absence condition.

(a)

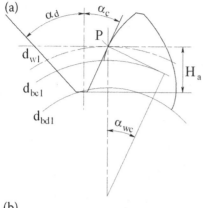

(b)

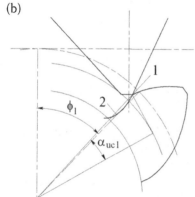

FIGURE 2.20
Definition of the undercut profile angle α_{uc1} in rack and pinion mesh; (a) initial tooth mesh position – rack and pinion teeth in contact at the pitch point P; (b) undercut position; 1: undercut profile point, 2: trajectory the mating rack tooth tip.

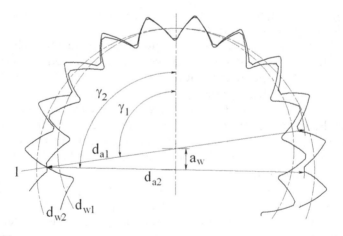

FIGURE 2.21
Tip/tip interference at the low pressure angle flanks in the internal mesh; 1: the interference point.

2.3.2.3 Transverse Contact Ratio of Asymmetric Gears

The transverse contact ratio is a length of a line of contact divided by the base pitch. In asymmetric tooth gears, the transverse contact ratios are defined separately for the drive and coast tooth flanks.

Figure 2.22 illustrates the definition of the transverse drive ε_{ad} and coast ε_{ac} contact ratios in the asymmetric external, internal, and rack and pinion meshes.

The transverse contact ratios are as follows:
for the external gear mesh, drive flanks:

$$\varepsilon_{ad} = \frac{z_1}{2\pi}(\tan \alpha_{ad1} + u \tan \alpha_{ad2} - (1 + u)\tan \alpha_{wd}), \qquad (2.124)$$

for the external gear mesh, coast flanks:

$$\varepsilon_{ac} = \frac{z_1}{2\pi}(\tan \alpha_{ac1} + u \tan \alpha_{ac2} - (1 + u)\tan \alpha_{wc}), \qquad (2.125)$$

for the internal gear mesh, drive flanks:

$$\varepsilon_{ad} = \frac{z_1}{2\pi}(\tan \alpha_{ad1} - u \tan \alpha_{ad2} + (u - 1)\tan \alpha_{wd}), \qquad (2.126)$$

for the internal gear mesh, coast flanks:

$$\varepsilon_{ac} = \frac{z_1}{2\pi}(\tan \alpha_{ac1} - u \tan \alpha_{ac2} + (u - 1)\tan \alpha_{wc}), \qquad (2.127)$$

for the rack and pinion mesh, drive flanks:

$$\varepsilon_{ad} = \frac{z_1}{2\pi}\left(\tan \alpha_{ad1} - \tan \alpha_d + \frac{2H_a}{d_{bd1} \sin \alpha_d}\right)$$

or

$$\varepsilon_{ad} = \frac{z_1}{2\pi}(\tan \alpha_{ad1} - \tan \alpha_d) + \frac{H_a}{P_{wd} \tan \alpha_d}. \qquad (2.128)$$

for the rack and pinion mesh, coast flanks:

$$\varepsilon_{ac} = \frac{z_1}{2\pi}\left(\tan \alpha_{ac1} - \tan \alpha_c + \frac{2H_a}{d_{bc1} \sin \alpha_c}\right)$$

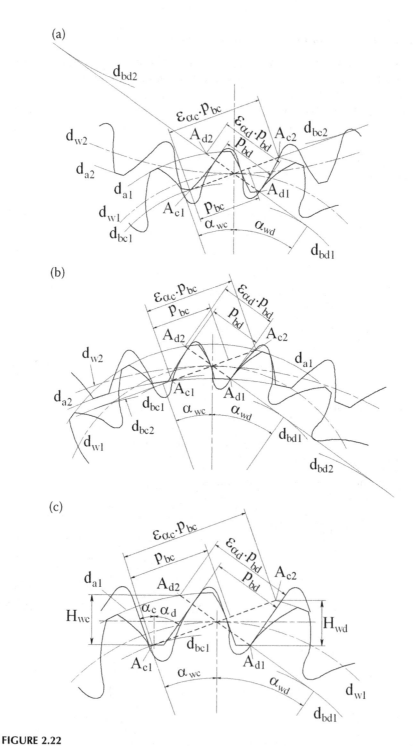

FIGURE 2.22
Transverse contact ratios in asymmetric gear mesh; A_{d1}–A_{d2} – drive flank line of contact, A_{c1}–A_{c2} – coast flank line of contact; (a) external, (b) internal, (c) rack and pinion.

or

$$\varepsilon_{\alpha c} = \frac{z_1}{2\pi}(\tan \alpha_{ac1} - \tan \alpha_c) + \frac{H_a}{p_{wc} \tan \alpha_c}. \tag{2.129}$$

Alternatively, the contact ratios for the external asymmetric gear meshes are defined with the following equations:
   for the drive flanks:

$$\varepsilon_{\alpha d} = \frac{z_1}{2\pi}((1 + u)\tan \alpha_{wd} - \tan \alpha_{pd1} - u \tan \alpha_{pd2}), \tag{2.130}$$

for the coast flanks:

$$\varepsilon_{\alpha c} = \frac{z_1}{2\pi}((1 + u)\tan \alpha_{wc} - \tan \alpha_{pc1} - u \tan \alpha_{pc2}). \tag{2.131}$$

The contact ratios for internal asymmetric gear meshes are defined with the following equations:
   for the drive flanks:

$$\varepsilon_{\alpha d} = \frac{z_1}{2\pi}(u \tan \alpha_{pd2} - \tan \alpha_{pd1} - (u - 1)\tan \alpha_{wd}), \tag{2.132}$$

for the coast flanks:

$$\varepsilon_{\alpha c} = \frac{z_1}{2\pi}(u \tan \alpha_{pc2} - \tan \alpha_{pc1} - (u - 1)\tan \alpha_{wc}). \tag{2.133}$$

The rack and pinion mesh contact ratios are defined with the following equations:
   for the drive flanks:

$$\varepsilon_{\alpha d} = \frac{z_1}{2\pi}\left(\tan \alpha_d - \tan \alpha_{pd1} + \frac{2(H_{wd} - H_{aw})}{d_{bd1} \sin \alpha_d}\right), \tag{2.134}$$

for the coast flanks:

$$\varepsilon_{\alpha c} = \frac{z_1}{2\pi}\left(\tan \alpha_c - \tan \alpha_{pc1} + \frac{2(H_{wc} - H_{aw})}{d_{bc1} \sin \alpha_c}\right), \tag{2.135}$$

where H_{wd} and H_{wc} are effective rack tooth depths.

Similarly to the gears with symmetric teeth, the transverse contact ratios of the gears with asymmetric teeth can be presented using parameters for only one of the mating gears:

for the drive flanks:

$$\varepsilon_{\alpha d} = \frac{z_1}{2\pi}(\tan \alpha_{ad1} - \tan \alpha_{pd1}), \tag{2.136}$$

$$\varepsilon_{\alpha d} = \frac{z_2}{2\pi}(\pm \tan \alpha_{ad2} \mp \tan \alpha_{pd2}), \tag{2.137}$$

for the coast flanks:

$$\varepsilon_{\alpha c} = \frac{z_1}{2\pi}(\tan \alpha_{ac1} - \tan \alpha_{pc1}), \tag{2.138}$$

$$\varepsilon_{\alpha c} = \frac{z_2}{2\pi}(\pm \tan \alpha_{ac2} \mp \tan \alpha_{pc2}), \tag{2.139}$$

where the top sign (+ or −) is for the external gears and the bottom sign (+ or −) is for the internal gears.

For a rack and pinion mesh the transverse contact ratios can be defined as follows:

for the drive flanks:

$$\varepsilon_{\alpha d} = \frac{H_{wd}}{p_{bd} \sin \alpha_d} = \frac{H_{wd}}{p_w \tan \alpha_d}, \tag{2.140}$$

for the coast flanks:

$$\varepsilon_{\alpha c} = \frac{H_{wc}}{p_{bc} \sin \alpha_c} = \frac{H_{wc}}{p_w \tan \alpha_c}. \tag{2.141}$$

In the metric system they are defined as:

for the drive flanks:

$$\varepsilon_{\alpha d} = \frac{H_{wd}}{\pi m_w \tan \alpha_d}, \tag{2.142}$$

for the coast flanks:

$$\varepsilon_{\alpha c} = \frac{H_{wc}}{\pi m_w \tan \alpha_c}. \tag{2.143}$$

In the English system they are:
for the drive flanks:

$$\varepsilon_{\alpha d} = \frac{DP \times H_{wd}}{\pi \tan \alpha_d},$$ (2.144)

for the coast flanks:

$$\varepsilon_{\alpha c} = \frac{DP \times H_{wc}}{\pi \tan \alpha_c}.$$ (2.145)

2.3.3 Helical Gearing Characteristics

A helical gear has theoretically correct involute tooth flank profiles only in the transverse section perpendicular to the gear axis. However, helical gears are typically specified by parameters in the normal to a tooth line section, because in many cases, they are fabricated using the generating rack tooling, for example, hob cutters. Parameters of this tooling are defined in the normal section and include a normal module or diametral pitch, normal pressure angle, and tooth proportions, like an addendum, whole depth, cutter tip radius, and an addendum modification (X-shift). Since the Direct Gear Design® method utilizes an established standard gear nomenclature and specifications as much as it is possible, helical gears are also specified with the normal section parameters.

In the metric system gear the normal module is

$$m_n = m_w \cdot \cos \beta_w,$$ (2.146)

In the English system gear the normal diametral pitch is

$$DP_n = DP_w / \cos \beta_w,$$ (2.147)

where β_w is a helix angle at the operating pitch diameters d_{w1} and d_{w2}.

2.3.3.1 Symmetric Gearing

The normal pressure angle is

$$\alpha_n = \arctan(\tan \alpha_w \cdot \cos \beta_w),$$ (2.148)

Spur gears have only a transverse contact ratio. Helical gears have also an axial contact ratio or overlap ratio that in addition to a transverse contact ratio results with a total contact ratio, which is as follows:

$$\varepsilon_\gamma = \varepsilon_\alpha + \varepsilon_\beta, \tag{2.149}$$

The axial contact ratio (or the overlap ratio) ε_β is defined by the angular shift ϕ of the helical gear front and back sections (see Figure 2.23)

$$\phi = \frac{2AC}{d_b} = \frac{2AB}{d_b} \tag{2.150}$$

or

$$\phi = \frac{2b_w}{d_b} \tan \beta_b, \tag{2.151}$$

where AC is the arc shift of the helical gear sections, AB is the shift of helical gear sections projected on the contact plane tangent to the base cylinder at the point A, b_w is contact face width, a distance between helical gear front and back sections in contact, and β_b is helix angle at the base cylinder.

For two mating helical gears

$$\phi_1 = u \times \phi_2. \tag{2.152}$$

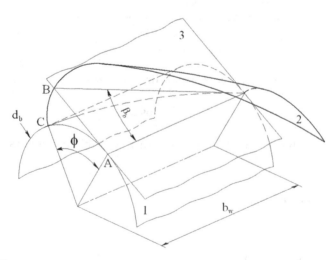

FIGURE 2.23
Angular shift ϕ of the helical gear front and back sections; 1: base cylinder, 2: helical involute surface, 3: contact plane tangent to the base cylinder.

The overlap ratio ε_β is defined by Equations (2.84), (2.150) or (2.151), and (2.152),

$$\varepsilon_\beta = \frac{AB}{p_b} = \frac{z_1}{2\pi}\phi_1 = \frac{z_2}{2\pi}\phi_2 = \frac{b_w}{p_b}\tan\beta_b. \qquad (2.153)$$

2.3.3.2 Asymmetric Gearing

The normal pressure angles of asymmetric gears are:
   for drive flanks:

$$\alpha_{nd} = \arctan(\tan\alpha_{wd}\cdot\cos\beta_w), \qquad (2.154)$$

for coast flanks:

$$\alpha_{nc} = \arctan(\tan\alpha_{wc}\cdot\cos\beta_w). \qquad (2.155)$$

The total contact ratio of asymmetric helical gears h is:
   for drive flanks:

$$\varepsilon_{\gamma d} = \varepsilon_{\alpha d} + \varepsilon_\beta, \qquad (2.156)$$

for coast flanks:

$$\varepsilon_{\gamma c} = \varepsilon_{\alpha c} + \varepsilon_\beta, \qquad (2.157)$$

For helical asymmetric tooth gears the drive and coast flank helix angles at the base cylinders are

$$\beta_{bd} = \arctan(\tan\beta \times \cos\alpha_{wd}) \qquad (2.158)$$

and

$$\beta_{bc} = \arctan(\tan\beta \times \cos\alpha_{wc}). \qquad (2.159)$$

The overlap ratio of asymmetric gears is identical for the drive and coast tooth flanks because according to Equations (2.28), (2.29), and (2.30)

$$p_{bc} = p_{bd}\frac{d_{bc}}{d_{bd}} = p_{bd} \times K \qquad (2.160)$$

and

$$\tan \beta_{bc} = \tan \beta_{bd} \frac{d_{bc}}{d_{bd}} = \tan \beta_{bd} \times K. \tag{2.161}$$

This allows presenting the overlap ratio Equation (2.153) for asymmetric gears as

$$\varepsilon_\beta = \varepsilon_{\beta d} = \varepsilon_{\beta c} = \frac{b_w}{p_{bd}} \tan \beta_{bd} = \frac{b_w}{p_{bc}} \tan \beta_{bc}. \tag{2.162}$$

2.4 Pitch Factor Analysis

This section presents an alternative method of involute gear geometry parameters and mesh definition, which is called the pitch factor analysis [27–29].

The gear mesh operating circular pitch is

$$p_w = \frac{\pi \times d_{w1,2}}{z_{1,2}} = S_{w1} + S_{w2} + S_{bl}, \tag{2.163}$$

where S_{w1} and S_{w2} are the pinion and gear tooth thicknesses, and S_{bl} is the arc backlash at the operating pitch diameter.

The tooth thicknesses S_{w1} and S_{w2} are (see Figures 2.24 and 2.25)
for symmetric gears:

$$S_{w1,2} = 2S_{1,2} + S_{v1,2}. \tag{2.164}$$

for asymmetric gears:

$$S_{w1,2} = S_{d1,2} + S_{c1,2} + S_{v1,2}, \tag{2.165}$$

where

$$S_{d1,2} = \frac{d_{w1,2}}{2} (\pm inv(\alpha_{ed1,2}) \mp inv(\alpha_{wd})) \tag{2.166}$$

are projections of the addendum portion of the drive involute flank on the pitch circle, and

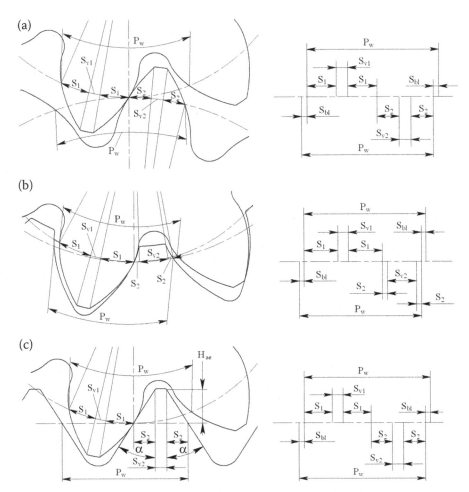

FIGURE 2.24
Symmetric gear mesh and operating pitch components; (a) external, (b) internal, (c) rack and pinion.

$$S_{c1,2} = \frac{d_{w1,2}}{2}\left(\pm inv\left(\alpha_{ec1,2}\right) \mp inv\left(\alpha_{wc}\right)\right) \tag{2.167}$$

are projections of the addendum portion of the coast involute flank on the pitch circle; the top sign (+ or −) is for the external gears and the bottom sign (+ or −) is for the internal gears; the $S_{v1,2}$ are the pitch circle projections of the tip land and radii.

For the rack and pinion gear mesh S_{d2}, S_{c2}, and S_{v2} are the pitch line projections. In this case, a projection of the addendum portion of the drive flank on the pitch line is

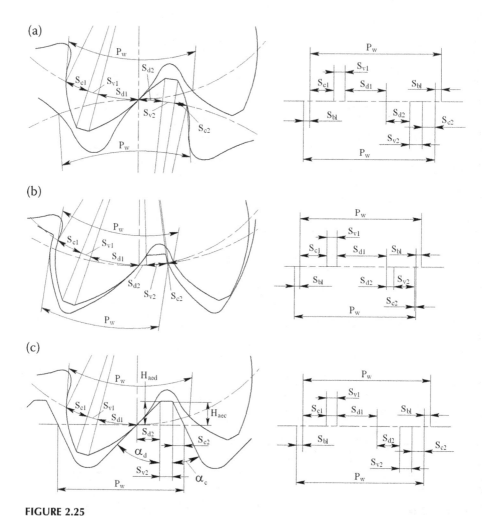

FIGURE 2.25
Asymmetric gear mesh and operating pitch components; (a) external, (b) internal, (c) rack and pinion.

$$S_{d2} = H_{aed} \tan \alpha_d. \tag{2.168}$$

A projection of the addendum portion of the coast flank on the pitch line is

$$S_{c2} = H_{aec} \tan \alpha_c. \tag{2.169}$$

Then the gear mesh operating circular pitch from Equation (2.163) is

$$p_w = S_{d1} + S_{d2} + S_{c1} + S_{c2} + S_{v1} + S_{v2} + S_{bl}. \tag{2.170}$$

A pitch factor equation is a result of division of Equation (2.170) by operating circular pitch p_w:

$$\theta_d + \theta_c + \theta_v = 1, \tag{2.171}$$

where θ_d is the drive pitch factor that is defined as

$$\theta_d = \frac{S_{d1} + S_{d2}}{p_w}, \tag{2.172}$$

θ_c is the coast pitch factor that is defined as

$$\theta_c = \frac{S_{c1} + S_{c2}}{p_w}, \tag{2.173}$$

and θ_v is the noncontact pitch factor that is defined as

$$\theta_v = \frac{S_{v1} + S_{v2} + S_{bl}}{p_w}. \tag{2.174}$$

The drive θ_d and coast θ_c pitch factors, as well as the asymmetry factor K, describe asymmetric gear geometry. However, the asymmetry factor K can be used to define both the gear tooth and mesh asymmetry, while the drive θ_d and coast θ_c pitch factors define only the gear mesh asymmetry.

The drive pitch factor is:
for the external gear mesh:

$$\theta_d = \frac{z_1}{2\pi}(inv\,(\alpha_{ed1}) + u\,inv\,(\alpha_{ed2}) - (1 + u)\,inv\,(\alpha_{wd})), \tag{2.175}$$

for the internal gear mesh:

$$\theta_d = \frac{z_1}{2\pi}(inv\,(\alpha_{ed1}) - u\,inv\,(\alpha_{ed2}) + (u - 1)\,inv\,(\alpha_{wd})). \tag{2.176}$$

The coast pitch factor is:
for the external gear mesh:

$$\theta_c = \frac{z_1}{2\pi}(inv\,(\alpha_{ec1}) - u\,inv\,(\alpha_{ec2}) + (u - 1)\,inv\,(\alpha_{wc})), \tag{2.177}$$

for the internal gear mesh:

$$\theta_c = \frac{z_1}{2\pi}(inv\,(\alpha_{ec1}) + u\,inv\,(\alpha_{ec2}) - (1 + u)\,inv\,(\alpha_{wc})). \tag{2.178}$$

For the rack and pinion gear mesh the drive and coast pitch factors are

$$\theta_d = \frac{z_1}{2\pi}(inv\,(\alpha_{ed1}) - inv\,(\alpha_d)) + \frac{H_{aed}}{p_w}\tan\alpha_d \qquad (2.179)$$

and

$$\theta_c = \frac{z_1}{2\pi}(inv\,(\alpha_{ec1}) - inv\,(\alpha_c)) + \frac{H_{aec}}{p_w}\tan\alpha_c. \qquad (2.180)$$

The drive and coast pressure angles are defined by equations:
  for the external gear mesh:

$$inv\,(\alpha_{wd}) = \frac{1}{1+u}\left(inv\,(\alpha_{ed1}) + u\,inv\,(\alpha_{ed2}) - \frac{2\pi\theta_d}{z_1}\right), \qquad (2.181)$$

$$inv\,(\alpha_{wc}) = \frac{1}{1+u}\left(inv\,(\alpha_{ec1}) + u\,inv\,(\alpha_{ec2}) - \frac{2\pi\theta_c}{z_1}\right), \qquad (2.182)$$

for the internal gear mesh:

$$inv\,(\alpha_{wd}) = \frac{1}{u-1}\left(\frac{2\pi\theta_d}{z_1} - inv\,(\alpha_{ed1}) + u\,inv\,(\alpha_{ed2})\right), \qquad (2.183)$$

$$inv\,(\alpha_{wc}) = \frac{1}{u-1}\left(\frac{2\pi\theta_c}{z_1} - inv\,(\alpha_{ec1}) + u\,inv\,(\alpha_{ec2})\right), \qquad (2.184)$$

for the rack and pinion gear mesh:

$$\alpha_{wd} = \alpha_d \qquad (2.185)$$

and

$$\alpha_{wc} = \alpha_c \qquad (2.186)$$

The external gear mesh transverse contact ratios are defined as following:
  as a solution of Equations (2.175) and (2.124) for the drive flanks:

$$\varepsilon_{ad} = \frac{z_1}{2\pi}\left(\alpha_{ed1} + u\alpha_{ed2} - (1+u)s\alpha_{wd} + \frac{2\pi\theta_d}{z_1}\right) \qquad (2.187)$$

and Equations (2.176) and (2.125) for the coast flanks:

$$\varepsilon_{ac} = \frac{z_1}{2\pi}\left(\alpha_{ec1} + u\alpha_{ec2} - (1+u)\alpha_{wc} + \frac{2\pi\theta_c}{z_1}\right). \tag{2.188}$$

The internal gear mesh transverse contact ratios are defined as following: as a solution of Equations (2.177) and (2.126) for the drive flanks:

$$\varepsilon_{ad} = \frac{z_1}{2\pi}\left(\alpha_{ed1} - u\alpha_{ed2} + (u-1)\alpha_{wd} + \frac{2\pi\theta_d}{z_1}\right) \tag{2.189}$$

and Equations (2.178) and (2.127) for the coast flanks:

$$\varepsilon_{ac} = \frac{z_1}{2\pi}\left(\alpha_{ec1} - u\alpha_{ec2} + (u-1)\alpha_{wc} + \frac{2\pi\theta_c}{z_1}\right) \tag{2.190}$$

In Equations (2.187–2.190), pressure angles and effective involute angles at the tooth tip are in radians.

The rack and pinion mesh drive and coast flank transverse contact ratios are defined by Equations (2.140) and (2.141).

For gears with symmetric teeth the angles $\alpha_{e1,2} = \alpha_{ed1,2} = \alpha_{ec1,2}$ and $\alpha_w = \alpha_{wd} = \alpha_{wc}$. Then the pitch factor $\theta = \theta_d = \theta_c$, that is:

for the external gear mesh:

$$\theta = \frac{z_1}{2\pi}(inv(\alpha_{e1}) + uinv(\alpha_{e2}) - (1+u)inv(\alpha_w))s, \tag{2.191}$$

for the internal gear mesh:

$$\theta = \frac{z_1}{2\pi}(inv(\alpha_{e1}) - uinv(\alpha_{e2}) + (u-1)inv(\alpha_w)), \tag{2.192}$$

for the rack and pinion gear mesh:

$$\theta = \frac{z_1}{2\pi}(inv(\alpha_{e1}) - inv(\alpha)) + \frac{H_{ae}}{P_w}\tan\alpha. \tag{2.193}$$

The pressure angle is defined by the following equations:

for the external gear mesh:

$$inv(\alpha_w) = \frac{1}{1+u}\left(inv(\alpha_{e1}) + uinv(\alpha_{e2}) - \frac{2\pi\theta}{z_1}\right), \tag{2.194}$$

for the internal gear mesh:

$$inv(\alpha_w) = \frac{1}{u-1}\left(\frac{2\pi\theta}{z_1} - inv(\alpha_{e1}) + u\,inv(\alpha_{e2})\right), \qquad (2.195)$$

for the rack and pinion gear mesh:

$$\alpha_w = \alpha. \qquad (2.196)$$

The external gear mesh transverse contact ratio is defined as a solution of Equations (2.191) and (2.85):

$$\varepsilon_\alpha = \frac{z_1}{2\pi}\left(\alpha_{e1} + u\alpha_{e2} - (1+u)\alpha_w + \frac{2\pi\theta}{z_1}\right). \qquad (2.197)$$

The internal gear mesh transverse contact ratio is defined as a solution of Equations (2.192) and (2.86):

$$\varepsilon_\alpha = \frac{z_1}{2\pi}\left(\alpha_{e1} - u\alpha_{e2} + \left(u-1\right)\alpha_w + \frac{2\pi\theta}{z_1}\right). \qquad (2.198)$$

In Equations (2.197–2.198), pressure angle and effective involute angles at the tooth tip are in radians.

The rack and pinion mesh transverse contact ratio is defined by Equation (2.93).

For gears with symmetric teeth the pitch factor θ from Equation (2.168) is

$$\theta = \theta_d = \theta_c = \frac{1}{2} \times (1 - \theta_v). \qquad (2.199)$$

This equation shows that for symmetric tooth gears, the pitch factor θ always ≤ 0.5. For the standard 20° pressure angle gears $\theta = 0.25$–0.30, and for the 25° pressure angle gears $\theta = 0.30$–0.35. For custom symmetric gears, the pitch θ can reach values of 0.40–0.45.

For gears with asymmetric teeth the drive pitch factor θ_d from Equation (2.171) is

$$\theta_d = 1 - \theta_c - \theta_v. \qquad (2.200)$$

Reduction of the coast pitch factor θ_c and the noncontact pitch factor θ_v allows the drive pitch factor θ_d to be significantly increased. Figure 2.26 presents the drive pressure angle versus the drive contact ratio $\alpha_{wd} - \varepsilon_{ad}$ chart at different values of θ_d for a gear couple with the pinion number of teeth $z_1 = 21$ and the gear number of teeth $z_2 = 37$. The chart shows that the

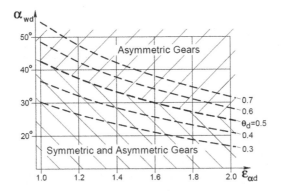

FIGURE 2.26
The $\alpha_{wd} - \varepsilon_{ad}$ chart with different values of the drive pitch factor θ_d for a gear couple with the pinion number of teeth $z_1 = 21$ and the gear number of teeth $z_2 = 37$.

symmetric gears lie below the curve $\theta_d = 0.5$. The asymmetric gears are located below and above this curve, which makes possible a simultaneous increase in the drive pressure angle and the drive contact ratio reducing the contact and bending stress, increasing load capacity, and power transmission density. This illustrates the potential advantages of the asymmetric gears over the symmetric ones for unidirectionally loaded gear drives.

Considering the theoretical case, when the noncontact pitch factor θ_v is equal to zero, the effective involute angles become equal to the respective profile angle at the intersection of the tooth flank involutes or $\alpha_{ed1,2} = \nu_{d1,2}$ and $\alpha_{ec1,2} = \nu_{c1,2}$, and a sum of the drive and coast pitch factors is equal one or $\theta_d + \theta_c = 1.0$ Then for the external asymmetric gears, a sum of Equations (2.181) and (2.182) becomes Equation (2.102). For the internal asymmetric gears, a sum of Equations (2.183) and (2.184) becomes Equation (2.103). Accordingly, for the external symmetric gears Equation (2.194) becomes Equation (2.67), and for the internal symmetric gears Equation (2.195) becomes Equation (2.68).

The pitch factor analysis is an additional Direct Gear Design® analytical tool that can be used for comparison of different gear geometry solutions, helping a gear designer better understand the available options and choose the optimal one.

2.5 Application of Direct Gear Design® for Different Types of Involute Gears

In the previous sections, the Direct Gear Design® approach was described mostly for the spur gears. However, it is also applicable to any other type of

TABLE 2.1

Virtual Spur Gear Conversion

Type of gear		Number of teeth of virtual spur gears
Spur	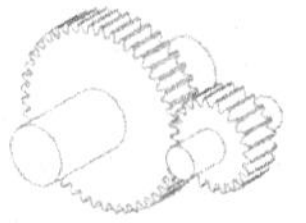	$Z_{1,2v} = Z_{1,2}$
Helical	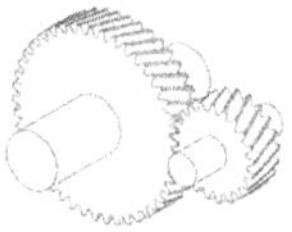	$Z_{1,2v} = \dfrac{Z_{1,2}}{\cos^3 \beta}$
Crossed Helical	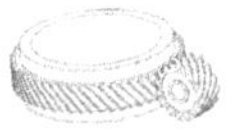	$Z_{1,2v} = \dfrac{Z_{1,2}}{\cos^3 \beta_{1,2}}$
Straight Tooth Bevel	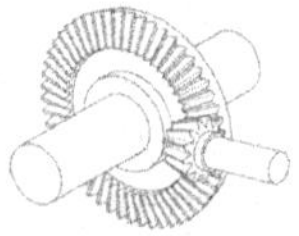	$Z_{1,2v} = \dfrac{Z_{1,2}}{\cos \gamma_{1,2}}$
Spiral and Skewed Bevel	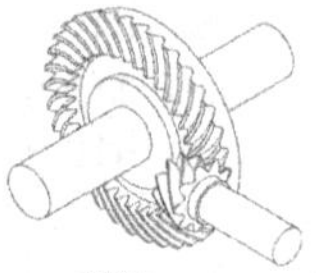 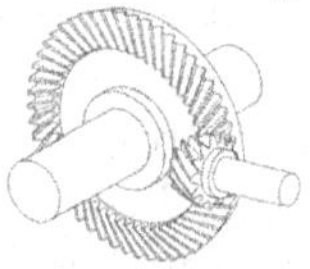	$Z_{1,2v} = \dfrac{Z_{1,2}}{\cos \gamma_{1,2} \cos^3 \beta}$
Hypoid	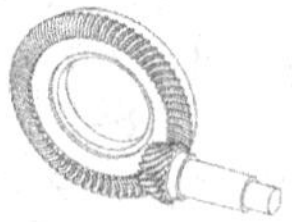	$Z_{1,2v} = \dfrac{Z_{1,2}}{\cos \gamma_{1,2} \cos^3 \beta_{1,2}}$
Worm	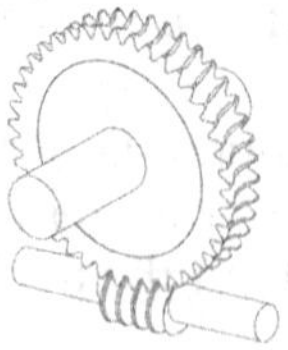	with involute worm: $Z_{wv} = Z_w / \sin^3 \beta$ $Z_{wgv} = Z_{wg} / \cos^3 \beta$ with Archimedes' worm: $Z_{wv} = \infty$ $Z_{wgv} = Z_w$
Face Spur		$Z_{1v} = Z_1$ $Z_{2v} = \infty$
Face Helical		$Z_{1v} = Z_1 / \cos^3 \beta$ $Z_{2v} = \infty$

(Continued)

TABLE 2.1 (Continued)

Type of gear		Number of teeth of virtual spur gears
Face Spiral		$Z_{1v} = Z_1/\cos^3 \beta_1$ $Z_{2v} = \infty$

Notes:
$Z_{1,2}$ – number of teeth of the real pinion and gear;
$Z_{1,2w}$ – number of teeth of virtual spur pinion and gear;
Z_w and Z_{wg} – number of starts of real worm and number of teeth of real worm gear;
Z_{wv} and Z_{wgv} – number of teeth of virtual gears that replace real worm and worm gear;
β – helix angle of helical or worm gears and spiral angle of spiral bevel gears;
$\gamma_{1,2}$ – pitch angle of bevel gears.

involute gears: helical, bevel, worm, face gears, etc. Tooth macrogeometry of these gears is typically defined in a normal plane to the tooth line. This normal plane tooth profile can be considered a tooth profile of the virtual spur gear. The formulas for calculating the number of teeth of virtual spur gears that have an identical tooth profile to the normal plane tooth profile of different types of involute types of gears are shown in Table 2.1. Virtual numbers of teeth are usually real numbers with the decimal parts. The asymmetric tooth geometry of virtual spur gears is optimized by means of Direct Gear Design® (see Chapter 5). Then the optimized tooth profiles are considered as the normal section tooth profiles of the actual gears.

3

Area of Existence of Involute Gears

3.1 Area of Existence Construction Conditions

Prof. E.B. Vulgakov had introduced the areas of existence of involute gears in his *Gear Theory in Generalized Parameters* [11]. He suggested using the profile angles $v_{1,2}$ at the intersection point of the tooth flank involutes as coordinates for an area of existence of two mating symmetric tooth gears with numbers of teeth z_1 and z_2. Some other gear tooth parameters also can be used as coordinates for an area of existence, for example, the tooth tip profile angles $\alpha_{a1,2}$ or the relative base tooth thicknesses $m_{b1,2}$ that are

$$m_{b1,2} = \frac{S_{b1,2}}{d_{b1,2}} = inv(v_{1,2}), \tag{3.1}$$

where $S_{b1,2}$ are base tooth thicknesses of the mating gears.

In this book, the profile angles $v_{1,2}$ at the intersection point of the tooth flank involutes are used as coordinates for an area of existence for symmetric tooth gears with numbers of teeth z_1 and z_2. For asymmetric tooth gears, coordinates for an area of existence are profile angles $v_{1,2d}$ at the intersection point of the drive flank involutes.

Areas of existence are described assuming the gear tooth tip radii and the gear mesh normal backlash equal to zero.

Unlike traditional design block contours, where the tooth thicknesses at the tooth tip diameters vary depending on the X-shifts, the area of existence is constructed for selected constant relative tooth tip thicknesses $m_{a1,2}$. In the book [11], the relative tooth tip thickness is defined as a ratio of the tip diameter tooth thicknesses to the base circle diameter:

$$m_{a1,2} = \frac{S_{a1,2}}{d_{b1,2}}. \tag{3.2}$$

However, it is more practical to present the relative tooth tip thickness ratio in relation to the operating module m_w (metric system) or the operating diametral pitch DP_w (English system), because such a ratio is commonly used to define a range of acceptable values for the tooth tip thickness. Then $m_{a1,2}$ can be described as

$$m_{a1,2} = \frac{S_{a1,2}}{m_w} \quad \text{or} \quad m_{a1,2} = S_{a1,2} \times DP_w. \tag{3.3}$$

For symmetric gears with external teeth the relative tooth tip thickness $m_{a1,2}$ is

$$m_{a1,2} = \frac{z_{1,2} \cos \alpha_w}{\cos \alpha_{a1,2}} (inv(v_{1,2}) - inv(\alpha_{a1,2})). \tag{3.4}$$

For symmetric gears with internal teeth the relative tooth tip thickness m_{a2} is

$$m_{a2} = \frac{z_2 \cos \alpha_w}{\cos \alpha_{a2}} \left(\frac{\pi}{z_2} - inv(v_2) + inv(\alpha_{a2}) \right). \tag{3.5}$$

For asymmetric gears with external teeth the relative tooth tip thickness $m_{a1,2}$ is

$$m_{a1,2} = \frac{z_{1,2} \cos \alpha_{wd}}{2 \cos \alpha_{ad1,2}} \times (inv(v_{d1,2}) + inv(v_{c1,2}) - inv(\alpha_{ad1,2}) - inv(\alpha_{ac1,2})). \tag{3.6}$$

For asymmetric gears with internal teeth the relative tooth tip thickness m_{a2} is

$$m_{a2} = \frac{z_2 \cos \alpha_{wd}}{2 \cos \alpha_{ad2}} \times (2\pi/z_2 - inv(v_{d2}) - inv(v_{c2}) + inv(\alpha_{ad2}) + inv(\alpha_{ac2})). \tag{3.7}$$

An area of existence contains a number of isograms, which represent the constant values of gear mesh parameters or the constant mesh conditions. The next sections of this chapter describe such isograms for gear pairs with the following parameters:

Pinion number of teeth $z_1 = 17$
Mating gear (external or internal) number of teeth $z_2 = 29$
Pinion relative tooth tip thickness $m_{a1} = 0.20$
Mating gear relative tooth tip thickness $m_{a1} = 0.30$

Asymmetry factors: $K = 1.0$ for symmetric gears and $K = 1.3$ for asymmetric gears

3.2 Pressure Angle Isograms

3.2.1 Symmetric Tooth Gears

Equations (2.67) and (2.68) define the pressure angle α_w = const isograms (Figure 3.1). These isograms do not depend on the relative tooth tip thicknesses $m_{a1,2}$.

The pressure angle isograms are bordered by isograms of the tooth tip profile angles α_{a1} = 0° and α_{a2} = 0°. These isograms are defined by Equations (2.67) and (3.4) for the external gears, and (2.68) and (3.5) for the internal gears.

The theoretical limits of the pressure angle are 0° < α_w < 90°. The practical range of the pressure angle varies depending on the type of involute gears and their application. For example, some worm gears with the metal worm and polymer worm gear have a very low (5–12°) pressure angle. The typical pressure angle range is 14.5–25°. High power density aerospace gears may have pressure angles 25–30°. Helical gears have both transverse and axial contact ratios (see Section 2.3.3). This allows realizing a significantly higher transverse pressure angle level. For example, the self-locking helical gears [32] can have transverse operating pressure angles up to 80° or higher.

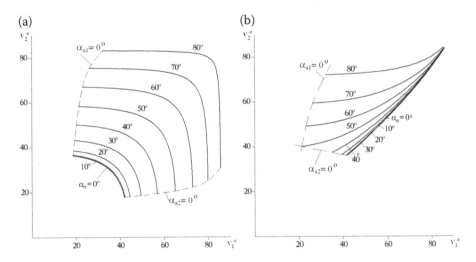

FIGURE 3.1
Pressure angle α_w = const isograms; (a) for external symmetric gears, (b) for internal symmetric gears.

3.2.2 Asymmetric Tooth Gears

The drive and coast pressure angle isograms α_{wd} = const and α_{wc} = const are defined by Equations (2.102) and (2.104) for the external gear pair (Figure 3.2a) and by Equations (2.103) and (2.104) for the internal gear pair (Figure 3.2b). Though the pressure angle isograms do not depend on the relative tooth tip thicknesses $m_{a1,2}$, they are bordered by isograms of the tooth tip profile angles $\alpha_{ac1} = 0°$ and $\alpha_{ac2} = 0°$, when the coast flanks become the points on the coast base diameters d_{bc1} and d_{bc2}. When the $\alpha_{ac1,2} = 0°$ the Equation (3.6) for external gears is

$$m_{a1,2} = z_{1,2} \cos \alpha_{wd} \times (inv(\nu_{d1,2}) + inv(\nu_{c1,2}) - inv(\arccos(1/K))). \quad (3.8)$$

When the $\alpha_{ac2} = 0°$ the Equation (3.7) for internal gears is

$$m_{a2} = z_2 \cos \alpha_{wd} \times (2\pi/z_2 - inv(\nu_{d2}) - inv(\nu_{c2}) + inv(\arccos(1/K))). \quad (3.9)$$

Then the isograms $\alpha_{ac1} = 0°$ and $\alpha_{ac2} = 0°$ are defined for external gears by Equations (2.102), (2.104), and (3.8), and for internal gears by Equations (2.102), (2.104), (3.8), and (3.9).

Minimal values of the pressure angles $\alpha_{wc} = 0°$ and $\alpha_{wd} = \arccos(1/K)$ occur when the gear pair pitch diameters become equal to the coast flank base diameters $d_{bc1,2}$.

The practical values of a drive flank pressure angle of asymmetric gears can be up to 45° and higher.

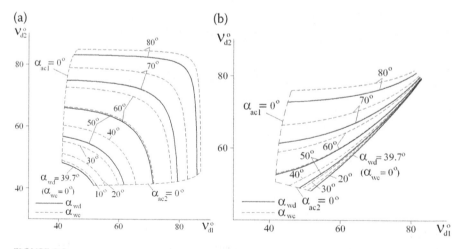

FIGURE 3.2
Pressure angle α_{wd} = const and α_{wc} = const isograms; (a) for external asymmetric gears, (b) for internal asymmetric gears.

3.3 Transverse Contact Ratio Isograms

3.3.1 Symmetric Tooth Gears

Figure 3.3a shows the transverse contact ratio ε_α = const isograms for the external gearing. They are described by the system of Equations (3.4) and

$$(1 + u) \times inv\left(arctan\left(\frac{1}{1+u}\left(tan\alpha_{a1} + u \times tan\alpha_{a2} - \frac{2\pi\varepsilon_\alpha}{z_1}\right)\right)\right)$$
$$- inv(v_1) - u \times inv(v_2) + \frac{\pi}{z_1} = 0, \quad (3.10)$$

which is a result of the combined solution of Equations (2.67) and (2.85).

The ε_α = const isograms for internal gearing (Figure 3.3b) are described by the system of Equations (3.4), (3.5), and

$$(u - 1) \times inv\left(arctan\left(\frac{1}{u-1}\left(u \times tan\alpha_{a2} - tan\alpha_{a1} + \frac{2\pi\varepsilon_\alpha}{z_1}\right)\right)\right)$$
$$- u \times inv(v_2) + inv(v_1) = 0, \quad (3.11)$$

which is a result of the combined solution of Equations (2.68) and (2.86).

Spur gears must have the contact ratio $\varepsilon_\alpha \geq 1.0$ to provide a smooth mesh transition from one pair of teeth to the following one. Isogram $\varepsilon_\alpha = 1.0$

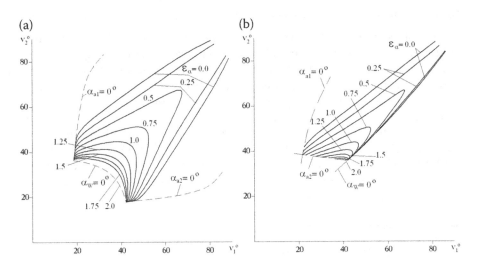

FIGURE 3.3
Transverse contact ratio ε_α = const isograms; (a) for external symmetric gears, (b) for internal symmetric gears.

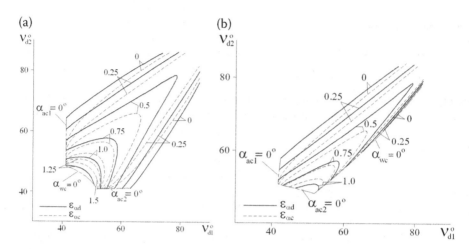

FIGURE 3.4

Transverse contact ratio ε_{ad} = const and ε_{ac} = const isograms; (a) for external asymmetric gears, (b) for internal asymmetric gears.

limits the selection of the spur gear combinations. Helical gears have an additional axial contact ratio (or overlap ratio) ε_β that allows the transverse contact ratio to be $\varepsilon_\alpha \geq 0$. The total contact ratio of helical gears must be $\varepsilon_\gamma = \varepsilon_\alpha + \varepsilon_\beta \geq 1.0$. Isogram $\varepsilon_\alpha = 0$ limits a choice of the helical gear combinations [33,34]. However, in most cases, helical gears have the transverse contact ratio $\varepsilon_\alpha \geq 1.0$.

3.3.2 Asymmetric Tooth Gears

Figure 3.4a shows the transverse contact ratio ε_{ad} = const and ε_{ac} = const isograms for the external gears. The drive flank contact ratio isograms ε_{ad} = const in this case are described as solutions of the Equations (2.102), (2.104), (2.124), and (3.6). The coast flank contact ratio isograms ε_{ac} = const are solutions of the Equations (2.102), (2.104), (2.125), and (3.6).

Figure 3.4b shows the transverse contact ratio ε_{ad} = const and ε_{ac} = const isograms for the internal gears. The drive flank contact ratio isograms ε_{ad} = const in this case are described as solutions of the Equations (2.103), (2.104), (2.126), (3.6), and (3.7). The coast flank contact ratio isograms ε_{ac} = const are solutions of the Equations (2.103), (2.104), (2.127), (3.6), and (3.7).

3.4 Overlap Ratio Isogram

An overlap ratio (axial contact ratio) can be presented as follows:
   For the metric system

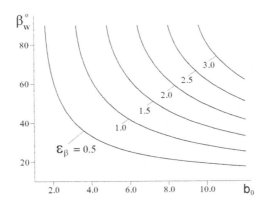

FIGURE 3.5
Overlap ratio ε_β = const isograms.

$$\varepsilon_\beta = \frac{b_w}{\pi \times m_w} \tan \beta_w = \frac{b_w}{\pi \times m_n} \sin \beta_w = \frac{b_0}{\pi} \sin \beta_{w'} \qquad (3.12)$$

For the English system

$$\varepsilon_\beta = \frac{b_w \times DP_w}{\pi} \tan \beta_w = \frac{b_w \times DP_n}{\pi} \sin \beta_w = \frac{b_0}{\pi} \sin \beta_{w'} \qquad (3.13)$$

where b_w: contact face width, m_w: operating module, m_n: normal module, DP_w: operating diametral pitch, DP_n: normal diametral pitch, β_w: helix angle at the operating pitch diameter, b_0: relative contact face width equal to b_w/m_n for the metric system or $b_w \times DP_n$ for the English system.

Figure 3.5 shows the overlap ratio ε_β = const isograms.

In helical gearing, it is preferable to have the integer overlap ratio ($\varepsilon_\beta = 1$, 2, 3, ...) that provides the minimal variation of a sum of the contact line lengths [7].

3.5 Interference Isograms

3.5.1 Symmetric Tooth Gears

Figure 3.6 shows the interference isograms.

Equations (2.69–2.72) define the profile angles α_{p1} and α_{p2} at the endpoints of the active involute flanks near the tooth root area. In the external gearing (Figure 3.6a) $\alpha_{p1} = 0°$ and $\alpha_{p2} = 0°$ describe a beginning of the root

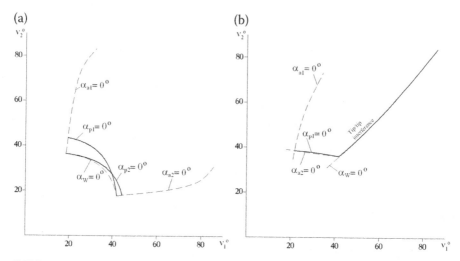

FIGURE 3.6
Interference isograms; (a) for external symmetric gears, (b) for internal symmetric gears.

interference (or undercut) for the pinion and the gear accordingly. The $\alpha_{p1} = 0°$ isogram is described by the system of Equations (3.4) and

$$(1 + u) \times inv\left(arctan\left(\frac{u}{1 + u} \times tan\,\alpha_{a2}\right)\right) - inv\,(\nu_1) - u \times inv\,(\nu_2) + \frac{\pi}{z_1} = 0,$$

$$(3.14)$$

which is a result of the combined solution of Equations (2.67) and (2.69). The $\alpha_{p2} = 0°$ isogram is described by the system of Equations (3.4) and

$$(1 + u) \times inv\left(arctan\left(\frac{1}{1 + u} \times tan\,\alpha_{a1}\right)\right) - inv\,(\nu_1) - u \times inv\,(\nu_2) + \frac{\pi}{z_1} = 0,$$

$$(3.15)$$

which is a result of the combined solution of Equations (2.67) and (2.70).

In the internal gearing (Figure 3.6b) the angle α_{p2} cannot be equal or close to zero, and only $\alpha_{p1} = 0°$ isogram describes the root area interference (or undercut) condition for the pinion. Its isogram is described by the system of Equations (3.5) and

$$(u - 1) \times inv\left(arctan\left(\frac{u}{u - 1} \times tan\,\alpha_{a2}\right)\right) + inv\,(\nu_1) - u \times inv\,(\nu_2) = 0,$$

$$(3.16)$$

which is a result of the combined solution of Equations (2.68) and (2.71).

The tip/tip interference (Section 2.3.1.3) is possible in the internal gear gearing. Its condition and isogram are defined by Equations (2.80–2.83) and (3.5).

Although the root interference isograms $\alpha_{p1} = 0°$ and $\alpha_{p2} = 0°$ for the external gearing, and $\alpha_{p1} = 0°$ for the external gearing present the borders of the area of existence, this does not mean that the gear meshes do not exist beyond these borders. However, those gear combinations have the tooth root undercut in at least one of the mating gears, and they are typically undesirable.

3.5.2 Asymmetric Tooth Gears

For asymmetric gears with an asymmetry factor $K > 1.0$, the coast tooth flank has lower involute profile angles than the drive one. As a result, an interference occurs first at the coast tooth flanks. The root interference isograms $\alpha_{pc1} = 0°$ and $\alpha_{pc2} = 0°$ for the external gearing reflect the coast flank root undercut conditions. They are defined as solutions of Equations (2.102), (2.106), (2.107), and (3.6) and are shown in Figure 3.7a.

The root interference $\alpha_{pc1} = 0°$ and tip/tip interference isograms for the internal gearing reflect the coast flank root undercut condition for the pinion and the tip/tip interference at the coast flanks. They are shown in Figure 3.7b. The $\alpha_{pc1} = 0°$ isogram is defined as a solution of Equations (2.103), (2.108), and (3.6). The tip/tip interference isogram is a solution of Equations (2.103), (2.108), (3.6), and (3.7).

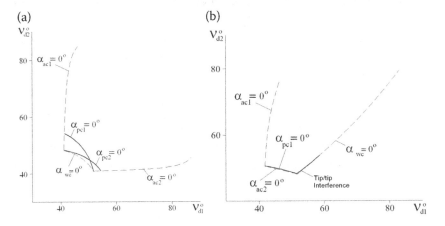

FIGURE 3.7
Interference isograms; (a) for external asymmetric gears, (b) for internal asymmetric gears.

3.6 Pitch Point Location Isograms

3.6.1 Symmetric Tooth Gears

In conventional gearing, the operating pitch diameter d_{w1} or d_{w2} divides the gear tooth height on the addendum and dedendum portions, and the pitch point P is located on the active part of the contact line $A1$–$A2$ (Figure 3.8). For external gears, this means that the operating pitch diameter is larger than the form diameter and smaller than the tooth tip diameter. For internal gears, this means that the operating pitch diameter is smaller than the form diameter and larger than the tooth tip diameter. Normally a gear mesh has the approach and recess actions while the contact point is moving along the contact line. An approach action is when the contact point C lies between point $A2$ and pitch point P of the contact line, or the driving pinion dedendum is in contact with the driven gear addendum. A recess action is when the contact point C lies between pitch point P and point $A1$ of the contact line, or the driving pinion addendum is in contact with the driven gear dedendum.

It is also possible to have only the approach action gearing (Figure 3.9) when the driving gear has the tip diameter $d_{a1} \leq\ \leq d_{w1}$ or $\alpha_{a1} \leq\ \leq \alpha_w$. In the approach action gearing, the driving gear tooth has only the dedendum without the addendum. Accordingly, the driven gear or rack tooth has only the addendum without dedendum. The pitch point P is located outside the active part of the contact line A_1–A_2 (beyond point A_1 in Figure 3.9).

The recess action gearing is shown in Figure 3.10. In the external mesh, the driven gear has the tooth tip diameter $d_{a2} \leq d_{w2}$ or $\alpha_{a2} \leq \alpha_w$. In the

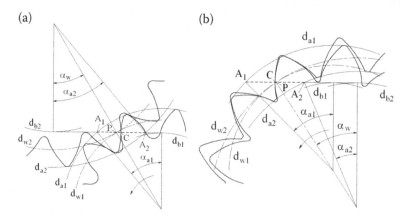

FIGURE 3.8
Conventional action gearing; (a) for external symmetric gears, (b) for internal symmetric gears.

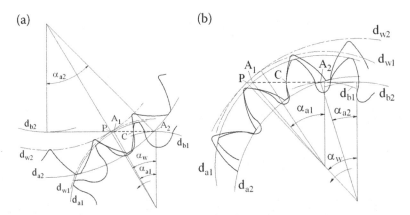

FIGURE 3.9
Approach action gearing; (a) for external symmetric gears, (b) for internal symmetric gears.

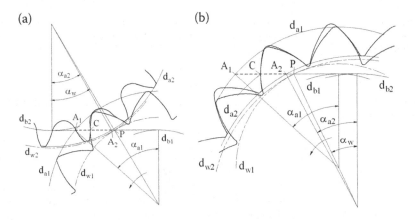

FIGURE 3.10
Recess action gearing; (a) for external symmetric gears, (b) for internal symmetric gears.

internal mesh, the driven gear has the tooth tip diameter $d_{a2} \geq d_{w2}$ or $\alpha_{a2} \geq \alpha_w$. In the recess action gearing, the driving gear tooth has only the addendum without the dedendum. Accordingly, the driven gear or rack tooth has only the dedendum without the addendum. The pitch point P is located outside the active part of contact line A_1–A_2 (beyond point A_2 in Figure 3.10).

The isograms of the pitch point location are shown in Figure 3.11. Isogram $\alpha_{a1} = \alpha_w$ is a border between the conventional and approach action gearing areas. Considering Equations (3.4) and (2.67), this isogram equation is

$$inv\,(\nu_1) - inv\,(\nu_2) + \frac{\pi - m_{a1}(1 + u)}{z_2} = 0. \qquad (3.17)$$

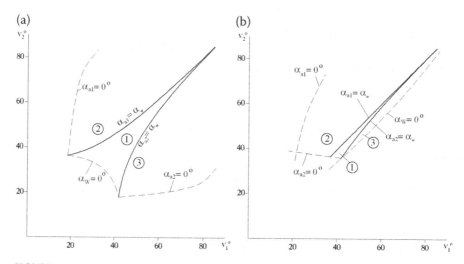

FIGURE 3.11
Pitch point location isograms; (a) for external symmetric gears, (b) for internal symmetric gears; 1: area of conventional action gearing; 2: area of the approach action gear; 3: area of the recess action gearing.

For the internal gearing, this isogram equation derives from the combined solution of Equations (3.4) and (2.68):

$$inv(\nu_1) - inv(\nu_2) - \frac{m_{a1}(u-1)}{z_2} = 0. \qquad (3.18)$$

Isogram $\alpha_{a2} = \alpha_w$ is a border between the conventional and recess action gearing areas. For external gearing, it is a solution of a system of Equations (3.4) and (2.67):

$$inv(\nu_1) - inv(\nu_2) + \frac{m_{a2}(1+u) - \pi u}{z_2} = 0. \qquad (3.19)$$

For the internal gearing, it is a solution of a system of Equations (3.5) and (2.68):

$$inv(\nu_1) - inv(\nu_2) - \frac{m_{a2}(u-1)}{z_2} = 0. \qquad (3.20)$$

The isograms of the pitch point location are situated between isograms $\alpha_{a1} = 0°$, $\alpha_{a2} = 0°$, and $\alpha_w = 0°$. Most gear applications use the conventional action gearing because it provides better performance parameters, such as high mesh efficiency (minimal tooth profile sliding), tooth surface durability, bending stress balance, etc. However, the approach and recess action gearings also may have rational areas of application. For example, the recess action gearing is used for the self-locking gears [32,34].

3.6.2 Asymmetric Tooth Gears

The definition of the pitch point location isograms for the asymmetric gears is similar to the symmetric ones. The conventional, approach, and recess action gearing schematics for external and internal gear pairs are shown in Figures 3.12, 3.13, and 3.14.

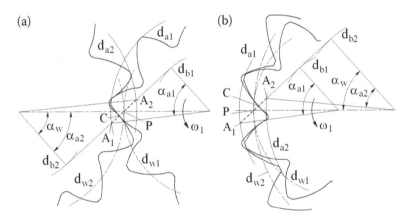

FIGURE 3.12
Conventional action gearing; (a) for external asymmetric gears, (b) for internal asymmetric gears.

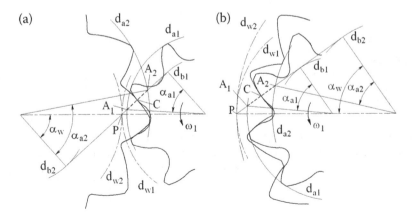

FIGURE 3.13
Approach action gearing; (a) for external asymmetric gears, (b) for internal asymmetric gears.

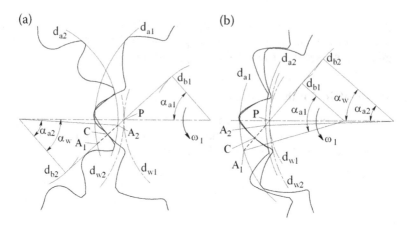

FIGURE 3.14
Recess action gearing; (a) for external asymmetric gears, (b) for internal asymmetric gears.

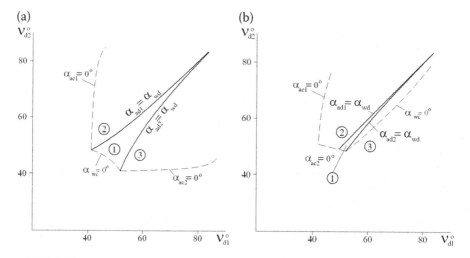

FIGURE 3.15
Pitch point location isograms; (a) for external asymmetric gears, (b) for internal asymmetric gears; 1: area of conventional action gearing; 2: area of the approach action gear; 3: area of the recess action gearing.

The isograms of the pitch point location of asymmetric gears are shown in Figure 3.15. Isogram $\alpha_{ad1} = \alpha_{wd}$ is a border between the conventional and approach action gearing areas. Considering Equations (3.6) and (2.102), this isogram equation is

$$inv(\nu_{d1}) + inv(\nu_{c1}) - inv(\nu_{d2}) - inv(\nu_{c2}) + \frac{2}{z_2} \times (\pi - m_{a1}(1 + u)) = 0.$$

$$(3.21)$$

For the internal gearing, this isogram equation derives from the combined solution of Equations (3.7) and (2.103):

$$inv(v_{d1}) + inv(v_{c1}) - inv(v_{d2}) - inv(v_{c2}) - \frac{2}{z_2} \times m_{a1}(u - 1) = 0. \quad (3.22)$$

Isogram $\alpha_{a2} = \alpha_w$ is a border between the conventional and recess action gearing areas. For external gearing, it is a solution of a system of Equations (3.6) and (2.102):

$$inv(v_{d1}) + inv(v_{c1}) - inv(v_{d2}) - inv(v_{c2}) + \frac{2}{z_2} \times (m_{a2}(1 + u) - \pi u) = 0.$$
$$(3.23)$$

For the internal gearing, it is a solution of a system of Equations (3.7) and (2.103):

$$inv(v_{d1}) + inv(v_{c1}) - inv(v_{d2}) - inv(v_{c2}) - \frac{2}{z_2} \times m_{a2}(u - 1) = 0. \quad (3.24)$$

3.7 Performance Parameters' Isograms

Many other isograms can be drawn in the area of existence. This section presents a few of them, which define gear pairs with certain constant performance characteristics. For asymmetric tooth gears these isograms are constructed for the drive tooth flanks in contact. In this section all equations are presented for asymmetric gears. However, they are valid for symmetric gears if by taking off the subscripted index "d." Figures are shown for asymmetric gears with the asymmetry factor being equal to 1.3.

Gear transmission power density and load capacity in many cases depend on the tooth surface durability. Maximization of gear transmission power density requires minimization of the contact (Hertz) stress σ_H. The minimal contact stress for the gear pairs with the constant operating pressure angle α_{wd} occurs at the maximum transverse contact ratio $\varepsilon_{\alpha d}$. It also happens for the gear pairs with the constant transverse contact ratio $\varepsilon_{\alpha d}$ when the operating pressure α_{wd} is maximum. These conditions occur for the drive flank of the asymmetric tooth gears when isograms $\alpha_{wd} = const$ and $\varepsilon_{\alpha d} = const$ are tangent (Figure 3.16).

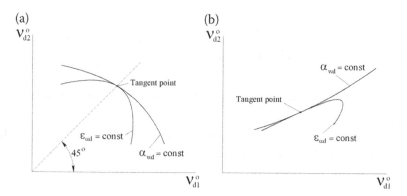

FIGURE 3.16
Tangent point of α_{wd} = const and ε_{ad} = const isograms: (a) for external gears; (b) for internal gears.

The σ_{Hmin} isograms (Figures 3.18 and 3.19) present the gear meshes that correspond to the tangent points of isograms α_{wd} = const and ε_{ad} = const. For the external asymmetric tooth gears, these isograms practically do not depend on the asymmetry factor K and are located approximately at 45° from the horizontal axis, where $\nu_{d1} \approx \nu_{d2}$.

The maximization of gear efficiency is critically important for many gear applications. Gear efficiency depends on gear mesh losses that for a pair of spur gears is defined as follows: [35]

for the drive flanks a pair of spur gears:

$$P_t = \frac{50f}{\cos \alpha_{wd}} \times \frac{H_{sd}^2 + H_{td}^2}{H_{sd} + H_{td}},\qquad(3.25)$$

for the drive flanks a pair of helical gears:

$$P_t = \frac{50f \cos^2 \beta}{\cos \alpha_{nd}} \times \frac{H_{sd}^2 + H_{td}^2}{H_{sd} + H_{td}},\qquad(3.26)$$

where f is an average friction coefficient, H_{sd} is a drive flank specific sliding velocity at the start of the approach action, and H_{td} is a drive flank specific sliding velocity at the end of the recess action.

Specific sliding velocities H_{sd} and H_{td} are ratios of the sliding velocity to the rolling velocity. They can be defined for the external and internal gears as follows:

$$H_{sd} = (u \pm 1) \times \cos \alpha_{wd} \times (\pm\tan \alpha_{ad2} \mp \tan \alpha_{wd}),\qquad(3.27)$$

$$H_{td} = \frac{u \pm 1}{u} \times \cos \alpha_{wd} \times (\tan \alpha_{ad1} - \tan \alpha_{wd}), \quad (3.28)$$

where the top sign (+ or −) is for the external gears and the bottom sign (+ or −) is for the internal gears.

Alternatively, from Equations (3.27), (2.106), or (2.107) for external gears, and Equations (3.28), (2.110), or (2.111) for internal gears, the specific sliding velocities are as follows:

$$H_{sd} = \frac{u \pm 1}{u} \times \cos \alpha_{wd} \times \left(\tan \alpha_{wd} - \tan \alpha_{pd1} \right), \quad (3.29)$$

$$H_{td} = (u \pm 1) \times \cos \alpha_{wd} \times \left(\pm \tan \alpha_{wd} \mp \tan \alpha_{pd2} \right). \quad (3.30)$$

From Equations (3.25) and (3.26), gear mesh losses reach their minimum when $H_{sd} = H_{td}$. It means that a condition for the maximum gear mesh efficiency E_{max} can be defined for the external and internal gears from (3.27) and (3.28) as follows:

$$\tan \alpha_{ad1} \mp u \tan \alpha_{ad2} \pm (u \mp 1) \tan \alpha_{wd} = 0, \quad (3.31)$$

where the top sign (+ or −) is for the external gears and the bottom sign (+ or −) is for the internal gears.

Equations of the maximum gear mesh efficiency E_{max} value isograms for symmetric gears (Figure 3.18) are defined as a solution of Equations (3.31) and (3.4) for the external gears, and Equations (3.31), (3.4), and (3.5) for the internal gears. Equations of the maximum gear mesh efficiency E_{max} value isograms for asymmetric gears (Figure 3.19) are defined as a solution of Equations (3.31) and (3.6) for the external gears, and Equations (3.31), (3.6), and (3.7) for the internal gears.

Gear tooth geometry, including the tooth flanks and root fillet, affects the maximum bending stress level. In traditional gear design, the tooth fillet profile is typically a trochoidal trajectory of the generating tool tooth tip. In Direct Gear Design®, the tooth flanks and root fillets are constructed independently, and the tooth fillet profile is optimized to minimize bending stress concentration (see Section 5.1.2). However, the tooth fillet profile optimization is a time-consuming process that is used for the final stage of gear design. It is not practical for browsing the area of existence, analyzing many sets of gear pairs with the optimized root fillets. For preliminary construction of the interference-free tooth root fillet profile that also provides relatively low bending stress concentration, the virtual ellipsis arc is built into the tooth tip that is tangent to the involute profiles at the tooth tip [36]. This makes the root fillet profile of one gear as a trajectory of the

(a)   (b)

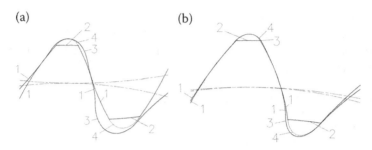

FIGURE 3.17
Tooth root fillet profile construction; (a) for external gears; (b) for internal gears; 1: involute profiles; 2: tooth tip lands; 3: fillet profiles; 4: ellipsis arcs.

mating gear tooth tip virtual ellipsis arc (see Figure 3.17). The virtual ellipsis arc is chosen because it seamlessly fits to an asymmetric tooth tip and results in a relatively low root bending stress level. This fillet profile can be considered pre-optimized because it provides lower bending stress concentration than the full tip radius rack generated fillet profile commonly used for bending stress reduction in the traditional gear design.

When the gear tooth with the root fillet is defined, the bending stress is calculated utilizing the finite element analysis (FEA) method.

If mating gears are made of similar materials and have a relatively close number of load cycles, the maximum bending stresses of mating gears should be equalized. The equal maximum bending stress isogram $\sigma_{F1max} = \sigma_{F2max}$ is shown in Figures 3.18 and 3.19. This isogram allows a preliminary selection of a pair of gears with the equalized bending strength. Later during final gear design, the mating gear face widths can be adjusted, considering also a number of load cycles for each gear to achieve more accurate bending strength equalization.

The equal maximum bending stress isogram is typical for the external gears. The internal gear tooth with the equal face width with its external mating gear usually has significantly lower bending stress because its root tooth thickness is typically much larger than that of the mating pinion. In this case, the bending stress balance can be achieved by the internal gear face width reduction.

3.8 Area of Existence and Gear Tooth Profiles

Summation of all the above-mentioned and other possible parameter or condition isograms forms the area of existence. Every point of the area of existence presents a gear pair with a certain set of parameters and gear tooth

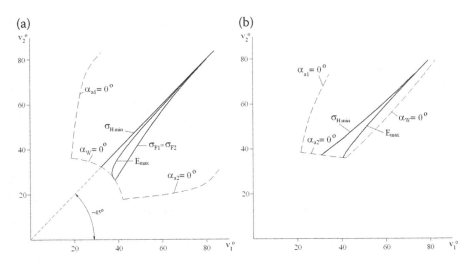

FIGURE 3.18
Isograms of performance parameters σ_{Hmin}, E_{max}, and $\sigma_{F1max} = \sigma_{F2max}$; (a) for external symmetric gears; (b) for internal symmetric gears.

profiles. Figures 3.20 and 3.21 present the areas of existence of the external and internal symmetric and asymmetric gears, and tooth profiles of the gear pairs corresponding to certain points of those areas. The shaded parts of the area of existence contain all reversible interference and tooth root undercut free gear pair options.

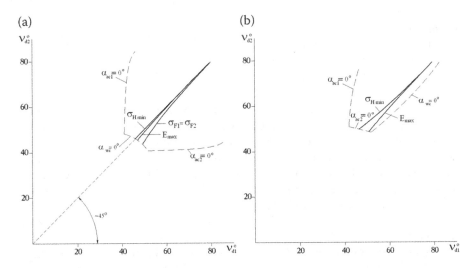

FIGURE 3.19
Isograms of performance parameters σ_{Hmin}, E_{max}, and $\sigma_{F1max} = \sigma_{F2max}$; (a) for external asymmetric gears; (b) for internal asymmetric gears.

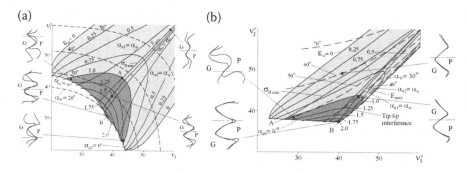

FIGURE 3.20
Area of existence and tooth profiles of the symmetric gears; (a) for external gears, (b) for internal gears; symbols P and G for the pinion and gear tooth profile, respectively.

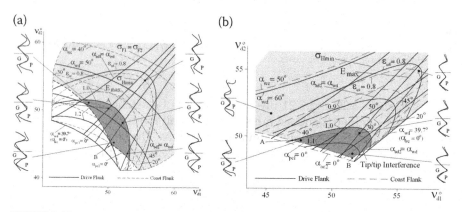

FIGURE 3.21
Area of existence and tooth profiles of the asymmetric gears; (a) for external gears, (b) for internal gears; symbols P and G for the pinion and gear tooth profile, respectively.

The darkly shaded portion of the area of existence includes the spur and helical gear pairs with the drive transverse contact ratio $\varepsilon_\alpha \geq 1.0$ for symmetric gears and $\varepsilon_{ad} \geq 1.0$ for asymmetric gears. The light shaded portion of the area of existence between the isograms $\varepsilon_\alpha = 1.0$ and $\varepsilon_\alpha = 0.0$ for symmetric gears, and the isograms $\varepsilon_{ad} = 1.0$ and $\varepsilon_{ac} = 0.0$ for asymmetric gears, includes only helical gear pairs. Existence of helical irreversible gear pairs is possible outside of the shaded areas between the isograms $\varepsilon_\alpha = 0.0$ and $\varepsilon_\alpha = 0.0$ for symmetric gears, and the isograms $\varepsilon_{ad} = 0.0$ and $\varepsilon_{ac} = 0.0$ for asymmetric gears.

More conventional or practical gear solutions are located within a darkly shaded portion with the drive contact ratio $\varepsilon_\alpha \geq 1.0$ for symmetric gears and $\varepsilon_{ad} \geq 1.0$ for asymmetric gears. Some gear pair tooth profiles have a kind of exotic shape and present rather theoretical interest. However, even such gear pairs may find practical applications for some unconventional gear drives.

A maximum drive pressure angle for the spur external gears is achieved in point A (Figures 3.20a and 3.21a). For symmetric gears at this point the pressure angle isogram $\alpha_w = $ const is tangent to the contact ratio isogram $\varepsilon_\alpha = 1.0$ (Figure 3.16a). For asymmetric gears in the point A the drive pressure angle isogram $\alpha_{wd} = $ const is also tangent to the drive contact ratio isogram $\varepsilon_{ad} = 1.0$, except for the gear pairs with high asymmetry factor K, high relative tooth tip thicknesses m_{a1} and m_{a2}, and a low number of teeth z_1 and z_2. In those cases, the point A is located in the intersection of the drive flank contact ratio isogram $\varepsilon_{ad} = 1.0$ and the coast flank interference isogram $\alpha_{pc1} = 0°$. For external symmetric gears, the minimum pressure angle and maximum contact ratio occur at point B at the intersection of the interference isograms $\alpha_{p1} = 0°$ and $\alpha_{p2} = 0°$. For external asymmetric gears, the minimum drive pressure angle and maximum drive contact ratio occur at point B at the intersection of the interference isograms $\alpha_{pc1} = 0°$ and $\alpha_{pc2} = 0°$.

In an area of existence of the spur internal gears, the point A (Figures 3.20b and 3.21b) presents the gear pair with the maximum pressure angle and contact ratio $\varepsilon_\alpha = 1.0$ for symmetric gears, and the gear pair with the maximum drive pressure angle and drive contact ratio $\varepsilon_{ad} = 1.0$ for symmetric gears. For symmetric gears the point A is located at the tangent point of the drive flank contact ratio isogram $\varepsilon_{ad} = 1.0$, and the drive pressure angle isogram $\alpha_{wd} = $ const (Figure 3.16b). For asymmetric gears, its location depends on the asymmetry factor K, relative tooth tip thicknesses m_{a1} and m_{a2}, and pinion and gear numbers of teeth z_1 and z_2. For a high asymmetry factor K, high relative tooth tip thicknesses m_{a1} and m_{a2}, and a low number of teeth z_1 and z_2, the point A lies in the intersection of the drive flank contact ratio isogram $\varepsilon_{ad} = 1.0$ and the coast flank interference isogram $\alpha_{pc1} = 0°$. Otherwise, it is located at the tangent point of the drive flank contact ratio isogram $\varepsilon_{ad} = 1.0$ and the drive pressure angle isogram $\alpha_{wd} = $ const (see Figure 3.16b). For symmetric gears, the point B is at the intersection of the interference isogram $\alpha_{p1} = 0°$ and the tip/tip interference isogram. For asymmetric gears the point B is at the intersection of the interference isogram $\alpha_{pc1} = 0°$ and the tip/tip interference isogram.

3.9 Area of Existence and Asymmetry Factors

The asymmetry factor K not only affects the shape of an asymmetric tooth (Figure 3.22) but also the size and shape of the area of existence. Figure 3.23

presents an overlay of the areas of existence of spur asymmetric tooth gears with different values of the asymmetric factor K. The largest area of existence is when the asymmetry factor K is equal to 1.0, and gear teeth are symmetric. This area is also located in a zone with relatively low flank profile angles. With the growth of an asymmetry factor K, gear teeth become more and more asymmetric, the area of existence gets smaller, and its location moves to a zone with greater flank profile angles. At the maximum spur gear value of the asymmetry factor K_{max}, the area of existence becomes a point, where the points A and B coincide. For external asymmetric gearing, this is a point of simultaneous intersection of the isogram of the drive flank contact ratio $\varepsilon_{ad} = 1.0$ and the coast flank interference isograms $\alpha_{pc1} = 0°$ and $\alpha_{pc2} = 0°$ (Figure 3.23a). For internal asymmetric gearing, this is a point of simultaneous intersection of the isogram of the drive flank contact ratio $\varepsilon_{ad} = 1.0$, the coast flank interference isogram $\alpha_{pc1} = 0°$, and tip/tip interference isogram (Figure 3.23b). If an asymmetry factor $K > K_{max}$, spur asymmetric gears are not possible because, in this case, the transverse drive flank contact ratio $\varepsilon_{ad} < 1.0$. However, this condition does not limit applications of helical asymmetric tooth gears with the total drive contact ratio $\varepsilon_{\gamma d} \geq 1.0$.

Figures 3.24 and 3.25 show how the pressure angle and contact ratio ranges are affected by the asymmetry factor K. These ranges are defined by differences in these parameter values between the points A and B of the area of existence. When $K = 1.0$, the gear teeth are symmetric, and the pressure angle and contact ratio ranges from their minimum to the maximum values are the largest. With the increasing asymmetry factor K, the pressure angle and contact ratio ranges are getting narrower. At $K = K_{max}$ these ranges are equal to zero, and points A and B coincide.

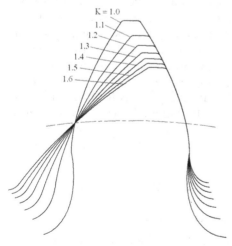

FIGURE 3.22
Tooth profiles with different asymmetry factors K, an identical coast flank pressure angle and tooth thicknesses at the pitch and tip diameters.

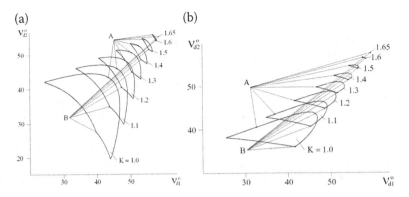

FIGURE 3.23
Areas of existence of the spur gears with different values of the asymmetry factor K; (a) for external gears, (b) for internal gears.

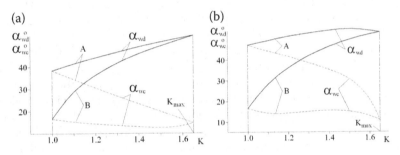

FIGURE 3.24
Pressure angle charts of the spur gears with different values of the asymmetry factor K; (a) for external gears, (b) for internal gears.

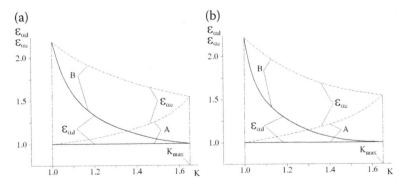

FIGURE 3.25
Contact ratio charts of the spur gears with different values of the asymmetry factor K; (a) for external gears, (b) for internal gears.

3.10 Area of Existence for Gear Pairs with Different Relative Tooth Tip Thicknesses

The selection of the relative tooth tip thicknesses m_{a1} and m_{a2} affects the size and shape of the area of existence. For asymmetric spur gears, values of relative tooth tip thicknesses can vary between theoretically zero and the maximum values $m_{a1,2max}$.

Figures 3.26 and 3.27 present an overlay of the areas of existence of the spur gears with different values of the relative tooth tip thicknesses m_{a1} and m_{a2}. The largest area of existence is when $m_{a1,2} = 0$ (shown in the dash thin lines) the gear teeth are pointed. This case presents only theoretical interest because the pointed gear teeth are totally impractical. For brittle gear materials, including the case hardened ones, the pointed tooth tip will be fractured under the load, and for more ductile materials, the pointed tooth tip will be bent. Both cases are unacceptable. The minimal value of the relative tooth tip thicknesses depends on gear materials, heat treatment, and applied load, and is typically equal to 0.2–0.3 for external gears and 0.3–0.4 for internal gears.

With the growth of the relative tooth tip thicknesses, the area of existence grows smaller, and its location changes toward a zone with lower tooth flank profile angles. At some maximum value of $m_{a1,2max}$, the area of existence reduces to a point. For external symmetric gearing (Figure 3.26a), this is a point of simultaneous intersection of the isogram of the contact ratio $\varepsilon_\alpha = 1.0$ and the interference isograms $\alpha_{p1} = 0°$ and $\alpha_{p2} = 0°$. For external asymmetric gearing (Figure 3.27a), this is a point of simultaneous

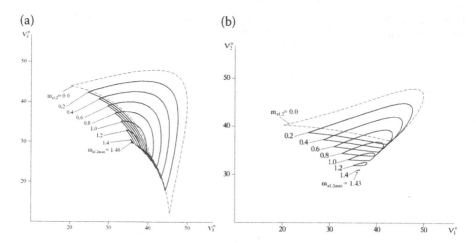

FIGURE 3.26
Areas of existence of the spur symmetric pinion and gear with $z_1 = 17$ and $z_2 = 29$, and different values of the relative tooth tip thicknesses $m_{a1,2}$; (a) for external gears, (b) for internal gears.

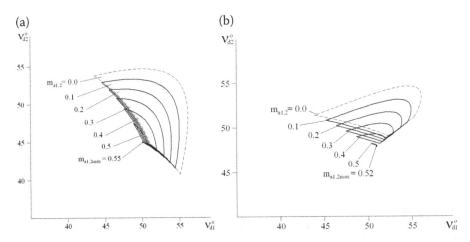

FIGURE 3.27
Areas of existence of the spur asymmetric pinion and gear with $z_1 = 17$ and $z_2 = 29$, $K = 1.3$, and different values of the relative tooth tip thicknesses $m_{a1,2}$; (a) for external gears, (b) for internal gears.

intersection of the isogram of the drive flank contact ratio $\varepsilon_{ad} = 1.0$ and the coast flank interference isograms $\alpha_{pc1} = 0°$ and $\alpha_{pc2} = 0°$. For internal symmetric gearing, this is a point of simultaneous intersection of the isogram of the contact ratio $\varepsilon_d = 1.0$, the interference isogram $\alpha_{p1} = 0°$, and the tip/tip interference isogram. For internal asymmetric gearing, this is a point of simultaneous intersection of the isogram of the drive flank contact ratio $\varepsilon_{ad} = 1.0$, the coast flank interference isogram $\alpha_{pc1} = 0°$, and the coast flank tip/ tip interference isogram. If the relative tooth tip thicknesses $m_{a1,2} > m_{a1,2max}$, spur gears are not possible, because, in this case, for the symmetric gearing the transverse contact ratio $\varepsilon_\alpha < 1.0$ and for the asymmetric gearing, the transverse drive flank contact ratio $\varepsilon_{ad} < 1.0$. However, this condition does not limit applications of helical gears with the total drive contact ratio greater than 1.0.

3.11 Area of Existence for Gear Pairs with Different Numbers of Teeth

Gear pairs with an identical gear ratio and different numbers of teeth have areas of existence that vary in sizes and shapes. Figures 3.28 and 3.29 show an overlay of areas of existence of the spur symmetric and asymmetric gears with different numbers of teeth and the given gear ratio, relative tooth tip thicknesses $m_{a1,2}$, and asymmetry factor K for asymmetric gears. The greater

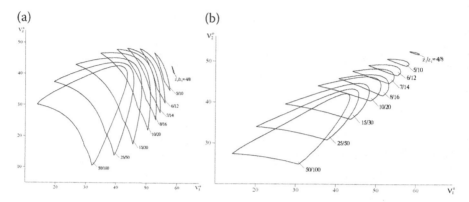

FIGURE 3.28
Areas of existence of the spur symmetric gears with $m_{a1,2} = 0.2$, $u = 2.0$, and different numbers of teeth; (a) for external gears, (b) for internal gears.

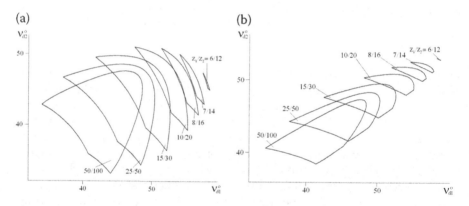

FIGURE 3.29
Areas of existence of the spur asymmetric gears with $K = 1.2$, $m_{a1,2} = 0.2$, $u = 2.0$, and different numbers of teeth; (a) for external gears, (b) for internal gears.

the number of teeth, the larger the area of existence. There are minimal numbers of the gear pair teeth when an area of existence becomes very small. An additional decrease in tooth numbers makes the spur gearing impossible because the transverse ratio becomes less than 1.0.

3.12 Area of Existence and Pitch Factors

In the previous sections of this chapter, the areas of existence of asymmetric gears have been considered with the given constant asymmetry factor K,

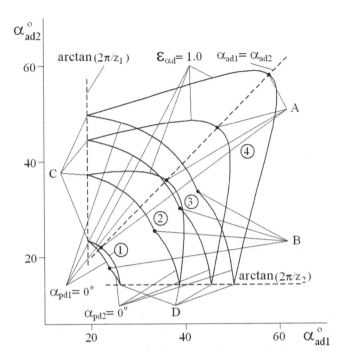

FIGURE 3.30
Areas of existence of the external spur gears with $z1 = 18$, $z2 = 25$ and different values of the pitch factor θ_d; 1: $\theta_d= 0.1$, 2: $\theta_d= 0.3$, 3: $\theta_d= 0.5$, 4: $\theta_d= 0.7$.

and the relative tooth tip thicknesses m_{a1} and m_{a2}. The pitch factors θ_d, θ_c, and θ_v (Section 2.4) in this case are varying. Figure 3.30 presents the overlaid areas of existence of spur external gears with the different constant drive flank pitch factors θ_d. This type of area of existence of involute gears defines only the drive flank gear meshes. If $\theta_d \leq 0.5$, the gears can have symmetric or asymmetric teeth. If $\theta_d > 0.5$, the spur gears can have only asymmetric teeth. The gears with symmetric teeth are always reversible. The gears with asymmetric teeth can be reversible or irreversible, depending on the coast flank pitch factor θ_c selection.

For the drive flanks of the external gears, the pressure angle isogram equation is defined from Equation (2.175), considering the tooth tip radii equal to zero. Then the effective tooth tip angles $\alpha_{ed1,2}$ should be replaced by the tooth tip angles $\alpha_{ad1,2}$:

$$inv\,(\alpha_{ad1}) + uinv\,(\alpha_{ad2}) - (1 + u)\,inv\,(\alpha_{wd}) - \frac{2\pi\theta_d}{z_1} = 0. \qquad (3.32)$$

The contact ratio isogram is defined by Equation (3.32) and the equation

$$\tan \alpha_{ad1} + u \tan \alpha_{ad2} - (1 + u)\tan \alpha_{wd} - \frac{2\pi\varepsilon_{ad}}{z_1} = 0. \qquad (3.33)$$

A result of subtracting Equation (3.32) from Equation (3.33) is

$$\alpha_{ad1} + u\alpha_{ad2} - (1 + u)\alpha_{wd} - \frac{2\pi(\varepsilon_{ad} - \theta_d)}{z_1} = 0. \qquad (3.34)$$

The interference isograms $\alpha_{pd1} = 0°$ and $\alpha_{pd2} = 0°$ are defined by Equation (3.32) and the equations

$$u \tan \alpha_{ad2} - (1 + u)\tan \alpha_{wd} = 0 \qquad (3.35)$$

and

$$\tan \alpha_{ad1} - (1 + u)\tan \alpha_{wd} = 0 \qquad (3.36)$$

that are derived from Equations (2.106) and (2.107).

In the point A, of an area of existence, where the drive flank pressure angle α_{wd} is maximum and the contact ratio $\varepsilon_{ad} = 1.0$, the pressure angle and contact ratio isograms have a common tangent point, and the first derivatives of these isogram functions should be equal:

$$\frac{d(\alpha_{ad2})}{d(\alpha_{ad1})}\bigg|_{\alpha_{wd}=const} = \frac{d(\alpha_{ad2})}{d(\alpha_{ad1})}\bigg|_{\varepsilon_{ad}=1.0} \qquad (3.37)$$

or with (3.32) and (3.33),

$$\frac{(\tan \alpha_{ad2})^2}{(\tan \alpha_{ad1})^2} = \frac{(\tan \alpha_{ad2})^2 + 1}{(\tan \alpha_{ad1})^2 + 1} \qquad (3.38)$$

or

$$\alpha_{ad1} = \alpha_{ad2}. \qquad (3.39)$$

It means the point A of areas of existence lies on the straight line $\alpha_{ad1} = \alpha_{ad2}$. The pressure angle equation at the point A is defined as a solution of Equations (3.32), (3.33), and (3.39):

$$\tan \alpha_{wd} + \frac{2\pi}{z_t} - \tan\left(\alpha_{wd} + \frac{2\pi(1 - \theta_d)}{z_t}\right) = 0, \qquad (3.40)$$

where $z_t = z_1 + z_2$ – total number of teeth of mating gears.

Its solution is [27]

$$\alpha_{wd}^{A} = \arctan\left(\sqrt{\frac{\pi^2}{z_t^2} + \frac{2\pi}{z_t \tan\frac{2\pi(1-\theta_d)}{z_t}} - 1} - \frac{\pi}{z_t}\right). \tag{3.41}$$

Then the coordinates of the point A at the area of existence are

$$\alpha_{ad1}^{A} = \alpha_{ad2}^{A} = \arctan\left(\sqrt{\frac{\pi^2}{z_t^2} + \frac{2\pi}{z_t \tan\frac{2\pi(1-\theta_d)}{z_t}} - 1} + \frac{\pi}{z_t}\right). \tag{3.42}$$

In point B, at the intersection of the interference isograms $\alpha_{pd1} = 0°$ and $\alpha_{pd2} = 0°$, the pressure angle is minimum, and the contact ratio is maximum. This maximum contact ratio is defined as a solution of Equations (3.33), (3.34), (3.35), and (3.36):

$$\arctan\left(\frac{2\pi\varepsilon_{ad}}{z_1}\right) + u\arctan\left(\frac{2\pi\varepsilon_{ad}}{z_2}\right) - (1+u)\arctan\left(\frac{2\pi\varepsilon_{ad}}{z_t}\right) - \frac{2\pi(\varepsilon_{ad} - \theta_d)}{z_1} = 0. \tag{3.43}$$

Then the coordinate angles α_{ad1} and α_{ad2}, and drive pressure angle α_{wd} at point B, are defined for drive flanks [34]:

$$\alpha_{ad1}^{B} = \arctan\left(\frac{2\pi\varepsilon_{ad}}{z_1}\right), \tag{3.44}$$

$$\alpha_{ad2}^{B} = \arctan\left(\frac{2\pi\varepsilon_{ad}}{z_2}\right), \tag{3.45}$$

and

$$\alpha_{wd}^{B} = \arctan\left(\frac{2\pi\varepsilon_{ad}}{z_t}\right). \tag{3.46}$$

The coordinate angle α_{ad1} at intersection point C of isograms $\varepsilon_{ad} = 1.0$ and $\alpha_{pd1} = 0°$ from Equations (3.34) and (3.35) is

$$\alpha_{ad1}^{C} = \arctan(2\pi/z_1). \tag{3.47}$$

The coordinate angle α_{ad2} and pressure angle α_{wd} at the point C are defined from equations

$$\arctan\left(\frac{2\pi}{z_1}\right) + u\alpha_{ad2} - (1 + u)\arctan\left(\frac{u}{1 + u}\tan\alpha_{ad2}\right) - \frac{2\pi(1 - \theta_d)}{z_1} = 0$$

(3.48)

and

$$\arctan\left(\frac{2\pi}{z_1}\right) + u\arctan\left(\frac{1 + u}{u}\tan\alpha_{wd}\right) - (1 + u)\alpha_{wd} - \frac{2\pi(1 - \theta_d)}{z_1} = 0.$$

(3.49)

The coordinate angle α_{ad2} at intersection point D of isograms $\varepsilon_{ad} = 1.0$ and $\alpha_{pd2} = 0°$ from Equations (3.34) and (3.36) is

$$\alpha_{ad2}^D = \arctan(2\pi/z_2).$$

(3.50)

The coordinate angle α_{ad1} and pressure angle α_{wd} at the point D are defined from equations

$$\alpha_{ad1} + u\arctan\left(\frac{2\pi}{z_2}\right) - (1 + u)\arctan\left(\frac{1}{1 + u}\tan\alpha_{ad1}\right) - \frac{2\pi(1 - \theta_d)}{z_1} = 0$$

(3.51)

and

$$\arctan\left((1 + u)\tan\alpha_{wd}\right) + u\arctan\left(\frac{2\pi}{z_2}\right) - (1 + u)\alpha_{wd} - \frac{2\pi(1 - \theta_d)}{z_1} = 0.$$

(3.52)

The pressure angle α_{wd}, contact ratio ε_{ad} charts at points A and B of the areas of existence from Figure 3.30 are presented in Figures 3.31 and 3.32.

Points of an area of existence with the constant drive flank pitch factor θ_d do not define complete mating gear teeth, but just their drive flanks. It allows independent selection of the tooth tip thicknesses and the coast tooth flank parameters of asymmetric gears. Considering the tooth tip radii and backlash equal to zero, the noncontact pitch factor θ_v from (2.174) is

$$\theta_v = \frac{m_{a1}\cos\alpha_{ad1} + m_{a2}\cos\alpha_{ad2}}{\pi\cos\alpha_{wd}}.$$

(3.53)

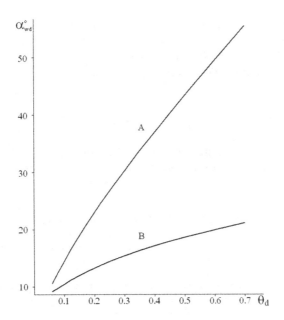

FIGURE 3.31
Drive pressure angle α_{wd} chart of the spur asymmetric gears with $z_1 = 18$, $z_2 = 25$ and different values of the drive pitch factor θ_d at the points A and B.

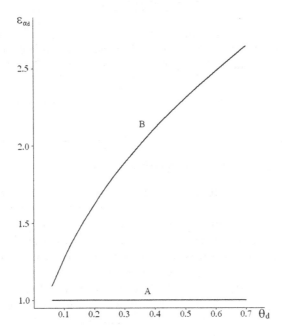

FIGURE 3.32
Drive contact ratio ε_{ad} chart of the spur asymmetric gears with $z_1 = 18$, $z_2 = 25$ and different values of the drive pitch factor θ_d at the points A and B.

When some point of the area of existence with coordinates α_{ad1} and α_{ad2} is chosen, the pressure angle α_{wd} is calculated by Equation (3.32). Then after selecting the relative tooth tip thicknesses m_{a1} and m_{a2}, the noncontact pitch factor θ_v is calculated by Equation (3.53). It allows defining the coast flank pitch factor as

$$\theta_c = 1 - \theta_d - \theta_v. \tag{3.54}$$

If the tooth tip radii are equal to zero that makes $\alpha_{ed1,2} = \alpha_{ad1,2}$ and $\alpha_{ec1,2} = \alpha_{ac1,2}$; the asymmetry factor K can be defined as a solution of Equations (2.104) and (2.177) as

$$(1 + u)inv\,(arc\,\cos(K\,\cos\,\alpha_{wd})) = inv\,(arc\,\cos(K\,\cos\,\alpha_{ad1}))$$
$$+ uinv\,(arc\,\cos(K\,\cos\,\alpha_{ad2})) - \frac{2\pi\theta_c}{z_1}. \tag{3.55}$$

Now the pressure angle and the coast flank tooth tip angles can be defined as

$$\alpha_{wc} = arc\cos(K\cos\alpha_{wd}), \tag{3.56}$$

$$\alpha_{ac1,2} = arc\cos(K\cos\alpha_{ad1,2}), \tag{3.57}$$

Equation (2.125) defines the coast flank contact ratio ε_{ac}. The profile angles at the points of contact near the root fillet of the drive and coast tooth flanks are described by Equations (2.106) and (2.107), and (2.108) and (2.109), respectively.

3.13 Application of Area of Existence

Initially, the areas of existence of involute gears were developed for only research purposes to learn more about capabilities and limitations of involute gear macrogeometry not limited by traditional gear design based on the gear rack generation. However, with the development of the area of existence calculation and construction software, it became one of the practically useful tools of Direct Gear Design®.

Figure 3.33 demonstrates an interface of the areas of existence calculation program. Input data include the following: pinion and gear numbers of teeth (for an internal gear pair, the gear number of teeth should be

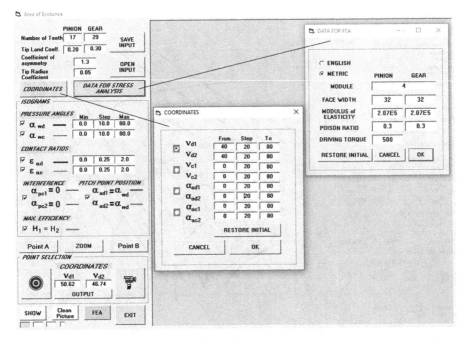

FIGURE 3.33
Area of existence program interface and input data.

shown a negative number), relative tooth tip thicknesses, also called tip land coefficients, asymmetry factor (equal to 1.0 for symmetric gears), and a tip radius coefficient (a ratio of the tooth tip radius to the operation module or diametral pitch) that, in this case, is identical for the pinion and gear, and the drive and coast tooth flank of asymmetric gears. The "*COORDINATES*" button opens a choice of an area of existence coordinates, including their initial minimum and maximum values, and incremental grid steps. The "*DATA FOR STRESS ANALYSIS*" button opens a window that contains a module for the metric system or diametral pitch for the English system, pinion and gear face widths, material properties including the modulus of elasticity and Poisson ratio, and pinion driving torque. The "*SAVE*" button allows storing selected data to open them later with the "*OPEN*" button.

The next step is the selection of isograms and their ranges. The "*SHOW*" button instantly shows the area of existence with selected isograms (Figure 3.34). Simultaneously, the program creates ASCII files with co-ordinate points of each selected isogram. Gear parameters of points *A* and *B* of the area of existence are shown by pressing the corresponding

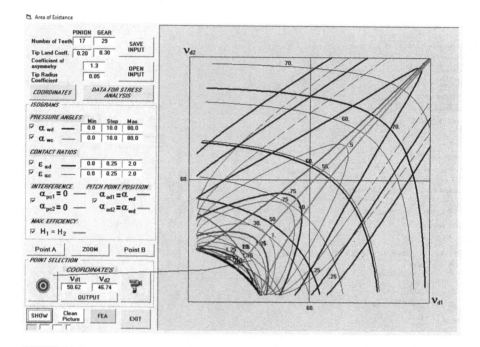

FIGURE 3.34

Area of existence isograms and the interface section presenting the gear pair data and coordinates of the selected point.

button. Any zone of the area of existence can be zoomed-in by the "ZOOM" button.

There are two ways to select some point of the area of existence: by a button with a target image or by typing the point coordinate values. A selected point of the area of existence presents a gear pair that could be animated by touching a button with the video camera image. It starts a gear mesh animation (Figure 3.35) that can be zoomed in or out and also speeded up or slowed down.

The "OUTPUT" button brings all geometry parameters of the selected gear pair to the interface screen (Figure 3.36) and creates the output file with these parameters. Such point selection allows quick evaluation of several points of the area of existence and chooses the most suitable one. Then pressing the "FEA" button brings the second program interface to calculate the selected gear pair stresses (Figure 3.37). On the bottom of the screen are all input data and the area of existence coordinates of the selected gear pair. The "RUN" button initiates the calculation of the Hertzian contact stress, bending stresses defined by

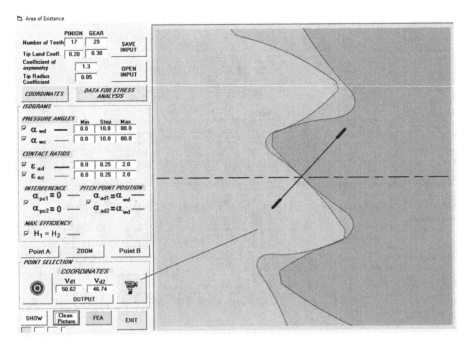

FIGURE 3.35
Animation of a gear pair, selected from area of existence.

the FEA, and gear mesh efficiency at the drive and flanks based on the chosen average friction coefficient.

An area of existence of involute gears can be used to find some unusual gear mesh solutions that were not known before. It also allows gear pairs with the required characteristics to be located. Its practical purpose is to define the gear pair parameters that satisfy specific performance requirements before detailed design and calculations. This involute gear research tool is incorporated into the preliminary design program with the FEA subroutine. Such a program can generate all isograms for gears with the given numbers of teeth $z1$ and $z2$, relative tooth tip thicknesses $ma1$ and $ma2$, and asymmetry factor K. It also calculates the relative sliding velocities, gear mesh efficiency, and bending and Hertzian contact stresses, as well as creating the ASCII tooth profiles and gear mesh animation of any point of the area of existence [34]. This software quickly defines limits of parameter selection of involute gears, locates and animates feasible gear pairs, and allows reviewing of their

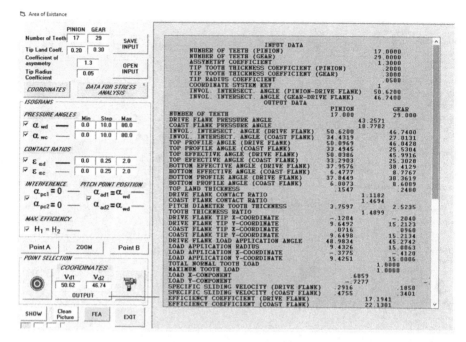

FIGURE 3.36
Output geometric parameters of a gear pair, selected from area of existence.

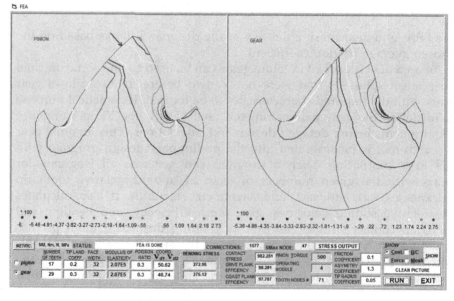

FIGURE 3.37
Stress analysis interface and calculation results.

geometry and stresses. Benefits of an area of existence include the following:

- consideration of all possible gear combinations;
- instant definition of limits of the gear performance parameters;
- awareness about some uncommon gear design options;
- quick localization of gear sets suitable for a particular application;
- preliminary gear design optimization.

4

Involute Gearing Limits

4.1 Numbers of Teeth

The selection of the number of teeth of the mating gears is critically important. First of all, it provides the required gear ratio. Second, when the gear ratio and center distance are specified, the selected numbers of teeth define the gear tooth size that is described by a module in the metric system or a diametral pitch in the English system. A gear tooth size is the main parameter in the definition of root bending stress. Third, the number of gear teeth is a major factor in the definition of gear mesh efficiency. Along with other gear mesh geometry parameters (pressure angles, contact ratios, etc.), tooth number selection allows the safety factors to be balanced for bending and contact stresses, and wear resistance to optimize a gear pair design. The maximum number of gear teeth is limited by application practicality and manufacturing technology. For example, most mechanically controlled gear hobbing machines can produce gears with up to 400 teeth using one-start hobs. The usage of multi-start hobs increases this limit. Some computer numerical control (CNC) gear hobbing machines can produce gears with up to 1,000 teeth using one-start hobs. Other gear fabrication technologies like, for example, profile cutting or injection molding can provide gears with an even greater number of teeth.

From an application point of view, the maximum number of gear teeth may also be limited by tolerance sensitivity and operating conditions. For a given gear pitch diameter, an increase in the number of teeth leads to tooth size reduction, to a point when the size of the tooth becomes comparable with tolerances achievable by selected gear fabrication technology. Gear drive operating conditions may also result in a similar effect for fine pitch gears. For example, a wide operating temperature range and application of dissimilar gear and housing materials may lead to noticeable gear size and center distance changes that could be comparable to the tooth size of the fine pitch gears. It may lead to a proper gear engagement disruption from gear jamming to complete tooth mesh separation.

4.1.1 Symmetric Tooth Gears

Gear handbooks give a conservative minimum number of teeth to avoid tooth root undercut by the tooling generating rack and reduce tooth flank sliding. For example, the gear handbook [35] indicates that tooth root undercut occurs for the 14½° pressure angle spur gears with a number of teeth lower than 32, for the 20° pressure angle spur gears with a number of teeth lower than 17, and for the 25° pressure angle spur gears with a number of teeth lower than 12. These numbers of teeth are defined by the beginning of the undercut condition when the standard addendum coefficient $h_a = 1.0$ and the rack shift coefficient $X = 0$.

Figure 4.1 explains the minimum number of teeth definition to avoid the root undercut. The tooth root undercut occurs when the generating rack addendum trajectory line $A–A$ is below the tangent point N, where the normal to the rack profile at the pitch point touches the base diameter d_b.

Then the undercut avoidance condition is

$$\frac{z_{min}\, m}{2} \sin^2\alpha \geq (h_a - x)m, \tag{4.1}$$

where m is a module and α is the rack pressure angle. From here

$$z_{min} \geq \frac{2(h_a - x)}{\sin^2\alpha}. \tag{4.2}$$

Application of the rack shift coefficient $X > 0$ or the nonstandard generating rack with high-pressure angle α and low addendum coefficient h_a allows

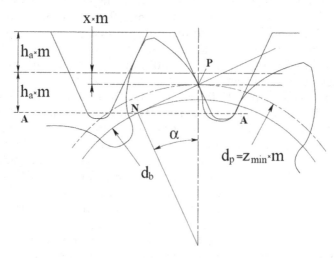

FIGURE 4.1
Definition of the minimum number of gear teeth to avoid the root undercut.

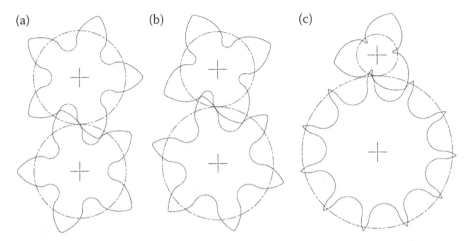

FIGURE 4.2
External symmetric spur gears with low number of teeth, (a) $z_{1,2} = 5$, (b) $z_1 = 4$, $z_2 = 6$, (c) $z_1 = 3$, $z_2 = 11$.

achieving a significantly lower minimum tooth number without tooth root undercut. Spur gears with a number of teeth as low as 6 are used in external gear pumps. Directly designed spur symmetric gears are not constrained by limitations imposed by the generating rack and its X-shift. Minimum numbers of teeth of external spur gears depend on the gear ratio and are defined by the simultaneous conditions: the involute profile angles at the lowest contact points $\alpha_{p1,2} = 0°$ (point B of an area of existence) and the contact ratio $\varepsilon_\alpha = 1.0$. The book [12] described a minimum number of teeth of external spur gears (Figure 4.2). The parameters of these gear pairs are presented in Table 4.1.

The pinion and gear tooth tip profile angles and the pressure angle are defined from Equations (3.47), (3.48), and (3.49), presented for symmetric gears, when $\varepsilon_\alpha = 1.0$. Then the flank pitch factor θ is from Equation (2.191), and non-contact pitch factor θ_v can be defined from (2.199) as

TABLE 4.1

External Symmetric Spur Gears with Low Number of Teeth

Pinion number of teeth, z_1	5	4	3
Gear number of teeth, z_2	5	6	11
Flank pitch factor, θ	0.46	0.48	0.495
Relative tooth tip thickness, $m_{a1,2}$	0.17	0.09	0.02
Pressure angle, $\alpha_w°$	32.07	32.1	24.19
Contact ratio, ε_α	1.0	1.0	1.0
Pinion tooth tip profile angle, $\alpha_{a1}°$	51.41	57.47	64.5
Gear tooth tip profile angle, $\alpha_{a2}°$	51.41	46.27	29.76

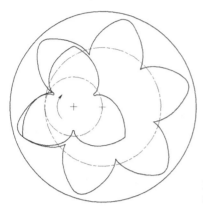

FIGURE 4.3
Internal symmetric spur gears with minimum number of teeth $z_1 = 3$, $z_2 = 6$.

$$\theta_v = 1 - 2\theta. \tag{4.3}$$

Assuming equal relative tooth tip thicknesses $m_{a1} = m_{a2}$, then using Equation (3.56) presented for symmetric gears:

$$m_{a1} = m_{a2} = \frac{\pi\theta_v \cos\alpha_w}{\cos\alpha_{a1} + \cos\alpha_{a2}}. \tag{4.4}$$

The following simultaneous conditions define the minimum numbers of teeth of internal spur gears: the pinion lowest contact point angle $\alpha_{p1} = 0°$, the beginning of the tip/tip interference (point B of the area of existence), and the contact ratio $\varepsilon_\alpha = 1.0$. Figure 4.3 shows the internal spur gears with minimum numbers of teeth. Its parameters are in Table 4.2.

Figure 4.4 presents an unusual epicyclic symmetric spur gear stage with a one-tooth sun gear, two one-tooth planet gears, and a three-tooth ring gear. The operation of this gear stage is also unusual. In a conventional epicyclic gear stage, all planet gears transmitting motion from the sun gear to the ring gear are simultaneously engaged with both of them. If this case, in the ring position angle = 0° (Figure 4.4a), the sun gear is engaged with both planet gears, and they are engaged with the ring gear. Then while moving from position 0° to position +90° (Figure 4.4b), the sun gear is driving only the left

TABLE 4.2

Internal Symmetric Spur Gears with Minimum Number of Teeth

Gear	External	Internal
Number of teeth, z	3	6
Transverse pressure angle, $\alpha_w°$	33.62	
Transverse contact ratio, ε_α	1.0	
Tooth tip profile angle, $\alpha_a°$	64.5	18.4

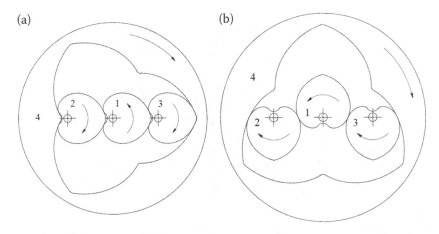

FIGURE 4.4
Epicyclic symmetric spur gear stage with minimum number of teeth: (a) ring position angle = 0°; (b) ring position angle = +90°; 1: sun gear; 2: left planet gear; 3: right planet gear; 4: ring gear.

planet gear, which transmits motion to the ring gear. The right planet gear does not transmit motion, but is moved by the ring gear. When the gears pass the ring position angle = +180°, the left and right planet gears' motion transmission roles are changed, i.e., during one half of the revolution of the sun gear, one planet gear transmits motion, then another one. This allows transmission of motion despite the contact ratios in both sun/planet and planet/ring gear meshes being lower than 1.0. Table 4.3 shows the parameters of such an epicyclic gear stage.

Helical gears also have an axial contact ratio ε_β that compensates for the lack of a transverse contact ratio ε_α (see Section 2.3.3), which can be reduced to zero. This makes it possible to have a minimum number of teeth $z_1 = z_2 = 1$ [27,34]. Figure 4.5 shows specimens of the external helical symmetric gears with numbers of teeth equaling 1 and 2. The main parameters of these gears are in Table 4.4. Figure 4.6 presents their areas of existence.

Figure 4.7 shows the internal helical symmetric gears with a minimum number of teeth. The main parameters of these gears are in Table 4.5.

TABLE 4.3

Epicyclic Symmetric Spur Gears with Minimum Number of Teeth

Gear	Sun		Planet		Ring
Number of teeth, z	1		1		3
Transverse pressure angle, $\alpha_w°$		65		65	
Transverse contact ratio, ε_α		0.56		0.59	
Tooth tip profile angle, $\alpha_a°$	75.6		75.6		56.1

(a)   (b)

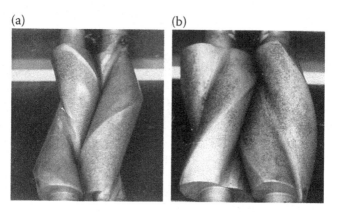

FIGURE 4.5
External symmetric helical gears with low number of teeth: (a) $z_{1,2} = 1$, (b) $z_{1,2} = 2$.

TABLE 4.4

External Symmetric Helical Gears with Low Number of Teeth

Number of teeth, $z_{1,2}$	1	2
Transverse pressure angle, $\alpha_w°$	65.0	65.3
Helix angle at pitch diameter, $\beta_w°$	29.2	29.5
Transverse contact ratio, ε_α	0.56	0.58
Axial contact ratio, ε_β	1.0	0.5
Total contact ratio, ε_γ	1.56	1.08
Tooth tip profile angle, $\alpha_{a1,2}°$	75.6	73.0

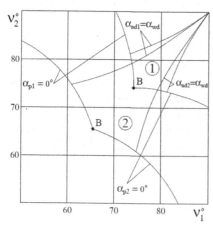

FIGURE 4.6
Areas of existence of external symmetric helical gears with minimum number of teeth: 1: $z_{1,2} = 1$; 2: $z_{1,2} = 2$.

4.1.2 Asymmetric Tooth Gears

Spur reversible asymmetric gears can transmit rotational motion and load by both drive and coast tooth flanks. It means that the drive and coast

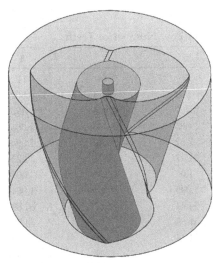

FIGURE 4.7
Internal symmetric helical gears with minimum number of teeth $z_1 = 1$, $z_2 = 2$.

contact ratios are $\varepsilon_{ad} \geq 1.0$ and $\varepsilon_{ac} \geq 1.0$. The selection of minimum numbers of teeth of the external spur reversible asymmetric gears depends on the gear ratio and is defined by the following simultaneous conditions: the coast flank profile angles at the points where involute flanks meet a root fillet $\alpha_{pc1,2} = 0°$ (point B of an area of existence) and the drive flank contact ratio $\varepsilon_{ad} = 1.0$. These minimum numbers of teeth are the same as for the external spur symmetric gears. The maximum asymmetry factor is achieved when the relative tip thickness $m_{a1,2} = 0.0$. This means that the noncontact pitch factor $\theta_v = 0.0$ and from Equation (2.171), the sum of the drive and coast pitch factors is equal to 1.0. Table 4.6 presents parameters of the external reversible asymmetric spur gears with a low number of teeth.

The pinion coast flank lowest contact point angle $\alpha_{pc1} = 0°$.

The following simultaneous conditions define the minimum numbers of teeth of internal reversible asymmetric spur gears: the pinion coast flank lowest contact point angle $\alpha_{pc1} = 0°$, the beginning of the coast

TABLE 4.5

Internal Symmetric Helical Gears with Minimum Number of Teeth

Gear	Pinion		Gear
Number of teeth	1		2
Transverse pressure angle, $\alpha_w°$		65.0	
Helix angle at pitch diameter, $\beta_w°$		61.0	
Transverse contact ratio, ε_α		0.61	
Transverse contact ratio, ε_β		0.5	
Tooth tip profile angle, $\alpha_{a1,2}°$	75.6		48.2

TABLE 4.6

External Reversible Asymmetric Spur Gears with Low Number of Teeth

Pinion number of teeth, z_1	5	4	3
Gear number of teeth, z_2	5	6	11
Drive flank pitch factor, θ_d	0.533	0.514	0.502
Coast flank pitch factor, θ_c	0.467	0.486	0.498
Asymmetry factor, K	1.068	1.030	1.006
Drive pressure angle, $\alpha_{wd}°$	37.62	34.80	24.94
Coast pressure angle, $\alpha_{wc}°$	32.27	32.23	24.25
Drive contact ratio, ε_{ad}	1.0	1.0	1.0
Coast contact ratio, ε_{ac}	1.005	1.004	1.004
Pinion tooth tip drive profile angle, $\alpha_{ad1}°$	54.44	58.67	64.69
Pinion tooth tip coast profile angle, $\alpha_{ac1}°$	51.63	57.61	64.57
Gear tooth tip drive profile angle, $\alpha_{ad2}°$	54.44	48.00	30.30
Gear tooth tip coast profile angle, $\alpha_{ac2}°$	51.63	46.42	29.83

flank tip/tip interference (point B of the area of existence), and the contact ratio $\varepsilon_{ad} = 1.0$. The maximum asymmetry factor is achieved when the relative tip thickness $m_{a1,2} = 0.0$. Table 4.7 provides parameters of the internal spur gears with minimum numbers of teeth. Because of a very low asymmetry factor ($K = 1.015$), the internal spur asymmetric gears with minimum numbers of teeth look practically the same as the internal spur symmetric gears with minimum numbers of teeth shown in Figure 4.3.

Figures 4.8 and 4.9 show the external and internal helical reversible asymmetric gears with a low number of teeth. Table 4.8 provides the main parameters of these gears.

TABLE 4.7

Internal Reversible Asymmetric Gears with Low Number of Teeth

Gear	External	Internal
Number of teeth, $z_{1,2}$	3	6
Drive flank pitch factor, θ_d	0.506	
Coast flank pitch factor, θ_c	0.494	
Asymmetry factor, K	1.015	
Drive pressure angle, $\alpha_{wd}°$	31.13	
Coast pressure angle, $\alpha_{wc}°$	29.67	
Drive contact ratio, ε_{ad}	1.0	
Coast contact ratio, ε_{ac}	1.009	
Drive tooth tip profile angle, $\alpha_{ad1,2}°$	65.16	56.95
Coast tooth tip profile angle, $\alpha_{ac1,2}°$	64.76	56.39

(a)   (b)

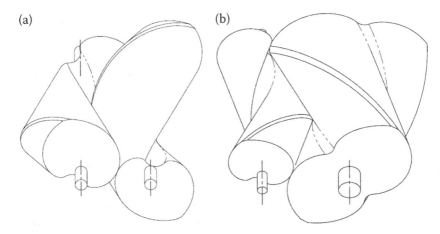

FIGURE 4.8
External helical reversible asymmetric gears with low number of teeth: (a) $z_{1,2} = 1$; (b) $z_1 = 1, z_2 = 2$.

The article [37] described the asymmetric spur gears with the number of teeth 1, 2, and 3, and gear ratio $u = 1$. Those gears have the drive contact ratio $\varepsilon_{ad} \geq 1.0$, but the coast contact ratio $\varepsilon_{ac} < 1.0$ makes them irreversible. Such gears may not have the coast flank involute profiles at all. Figure 4.10 shows the examples of these irreversible asymmetric spur gears with the number of teeth $z_{1,2} = 1$, $z_1 = 1$, and $z_2 = 2$. Table 4.9 presents the main parameters of these gears.

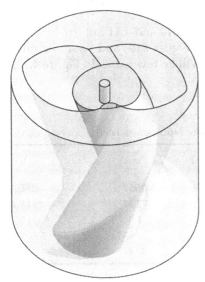

FIGURE 4.9
Internal helical reversible asymmetric gears with low number of teeth: $z_1 = 1, z_2 = 2$.

TABLE 4.8

Asymmetric Reversible Helical Gears with Low Number of Teeth

Gear mesh	External Figure 4.8a	External Figure 4.8b	Internal Figure 4.9
Pinion number of teeth, z_1	1	1	1
Gear number of teeth, z_2	1	2	2
Drive flank pitch factor, θ_d	0.594	0.584	0.519
Coast flank pitch factor, θ_c	0.368	0.358	0.303
Asymmetry factor, K	1.517	1.517	1.517
Drive pressure angle, $\alpha_{wd}°$	67.80	67.80	67.80
Coast pressure angle, $\alpha_{wc}°$	55.02	55.02	55.02
Helix angle at pitch diameter, $\beta_w°$	36.00	36.00	36.00
Drive transverse contact ratio, ε_{ad}	0.647	0.644	0.602
Coast transverse contact ratio, ε_{ac}	0.455	0.458	0.455
Axial contact ratio, ε_β	0.7	0.7	0.7
Drive total contact ratio, $\varepsilon_{\gamma d}$	1.347	1.344	1.302
Coast total contact ratio, $\varepsilon_{\gamma c}$	1.155	1.158	1.155
Pinion drive tooth tip profile angle, $\alpha_{ad1}°$	77.43	77.43	77.43
Gear drive tooth tip profile angle, $\alpha_{ad2}°$	77.43	73.87	53.60
Pinion coast tooth tip profile angle, $\alpha_{ac1}°$	54.83	54.83	54.83
Gear coast tooth tip profile angle, $\alpha_{ac2}°$	54.83	65.08	35.55

Figure 4.11 shows the internal irreversible asymmetric spur gears with the minimum number of teeth $z_1 = 2$, $z_2 = 5$. The tip/tip interference condition limits the minimal numbers of teeth of this internal gear mesh. Table 4.10 provides the main parameters of this gear pair.

Although the gear meshes shown in Figure 4.10 and 4.11 are irreversible, if they are assembled in two layers along the gear axes and have the mirrored tooth orientation, such a gear system will be reversible. In Figure 4.12,

TABLE 4.9

Irreversible External Asymmetric Gears with Low Number of Teeth

Pinion number of teeth, z_1	1	1
Gear number of teeth, z_2	1	2
Drive flank pitch factor, θ_d	0.952	0.909
Drive flank pressure angle, $\alpha_{wd}°$	72.34	64.50
Drive flank contact ratio, ε_{ad}	1.0	1.0
Pinion drive flank tooth tip profile angle, $\alpha_{ad1}°$	80.95	80.95
Gear drive flank tooth tip profile angle, $\alpha_{ad2}°$	80.95	72.31

TABLE 4.10

Irreversible Internal Asymmetric Spur Gears with Low Number of Teeth

Gear	External		Internal
Number of teeth, $z_{1,2}$	2		5
Drive flank pitch factor, θ_d		0.771	
Drive pressure angle, $\alpha_{wd}°$		65.04	
Drive contact ratio, ε_{ad}		1.0	
Drive tooth tip profile angle, $\alpha_{ad1,2}°$	72.31		56.95

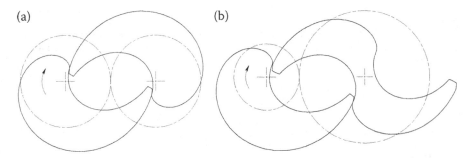

(a)   (b)

FIGURE 4.10
External spur irreversible asymmetric spur gears with low number of teeth: (a) $z_{1,2} = 1$; (b) $z_1 = 1$, and $z_2 = 2$.

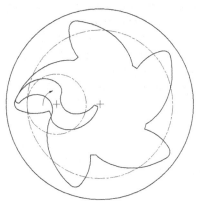

FIGURE 4.11
Internal irreversible asymmetric spur gears with minimum number of teeth $z_1 = 2$, $z_2 = 5$.

the front layer left driving gear operates in a counterclockwise direction, and the back layer gears are not engaged in motion transmission. In the reversed rotation, the back layer left driving gear operates in a clockwise direction, and the front layer gears are not engaged.

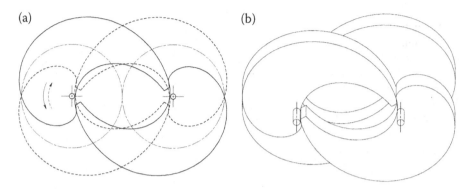

FIGURE 4.12
Two-layer assembly of reversible asymmetric spur gears with minimal number of teeth $z_{1,2} = 1$; (a) gear tooth profiles, front layer gear mesh – solid lines, back layer gear mesh – dashed lines; (b) isometric view.

4.2 Pressure Angles

The vast majority of gears are designed with standard tooth proportions. One of the main tooth proportion parameters is a pressure angle. Most gears are designed with the standard 20° pressure angle. The old standard pressure angle of 14½° is still in use. In some industries like, for example, aerospace, 25° and 28° pressure angles are used [8,38]. The term pressure angle, in this case, is related not to the gear mesh, but to the basic or generating rack that is used for design or as a cutter profile, accordingly. The gear (except the gear rack) involute profile angle varies from the form diameter to the tooth tip diameter. This section describes the transverse operating pressure angles, which are defined for a pair of mating gears as:

for symmetric gears:

$$\alpha_w = \arccos\left(\frac{d_{b1}(u \pm 1)}{2a_w}\right), \tag{4.5}$$

for asymmetric gears:

$$\alpha_{wd} = \arccos\left(\frac{d_{bd1}(u \pm 1)}{2a_w}\right) \text{ and } \alpha_{wc} = \arccos\left(\frac{d_{bc1}(u \pm 1)}{2a_w}\right). \tag{4.6}$$

For symmetric and asymmetric spur gears, the minimum pressure angle is defined at the point B of the area of existence intersection at the interference isograms $\alpha_{p1} = 0°$ and $\alpha_{p2} = 0°$ for symmetric gears and $\alpha_{pc1} = 0°$ and $\alpha_{pc2} = 0°$ for asymmetric gears. The higher the total number of teeth $z_t = z_1 + z_2$ and gear ratio u, the lower the pressure angle at point B.

For helical gears with symmetric teeth, the minimum theoretical pressure angle is $\alpha_w = 0°$. For helical gears with asymmetric teeth, the minimum theoretical pressure angles are $\alpha_{wc} = 0°$ and $\alpha_{wd} = \arccos(1/K)$.

The maximum pressure angle for external spur gears with symmetric teeth is defined at point A of the area of existence, where the pressure angle isogram is tangent to the transverse contact ratio isogram $\varepsilon_\alpha = 1.0$. It depends on the number of teeth of mating gears $z_{1,2}$ and also on the relative tooth tip thicknesses $m_{a1,2}$. The theoretical maximum pressure angle is achieved for a gear pair with the pointed teeth when $m_{a1,2} = 0$ and is defined as [12]

$$\alpha_{w\,max} = \arctan\left(\sqrt{\frac{\pi^2}{z_t^2} + \frac{2\pi}{z_t \tan\frac{\pi}{z_t}} - 1} - \frac{\pi}{z_t}\right). \tag{4.7}$$

According to Equation (4.7), when the total number of teeth is increasing and approaches infinity ($z_t \to \infty$), the pressure angle limit for external spur gears with symmetric teeth is $\alpha_{w\lim} = 45°$.

The general solution for the maximum pressure angle for both symmetric and asymmetric gears is presented in Equation (3.44). When the total number of teeth $z_t \to \infty$, the pressure angle limit for external spur gears from this equation is defined as [28]

$$\alpha_{wd}^{max\,lim} = \arctan\sqrt{\frac{\theta_d}{1 - \theta_d}}. \tag{4.8}$$

A chart of the maximum drive flank pressure angles α_{wdmax} for different total numbers of teeth z_t and the drive pitch factors θ_d is shown in Figure 4.13. If the drive pitch factor $\theta_d \leq 0.5$, this chart shows the maximum drive flank pressure angles for both symmetric and asymmetric gears; if the drive pitch factor $\theta_d > 0.5$, the maximum drive flank pressure angles are only for asymmetric gears.

The maximum pressure angle for external spur gears with asymmetric teeth is defined at point A of the area of existence, where the pressure angle isogram is tangent to the transverse contact ratio isogram $\varepsilon_{ad} = 1.0$. Figures 4.14 and 4.15 present the ranges of the drive and coast pressure angles of external spur asymmetric reversible gears with different numbers of teeth and asymmetry factors.

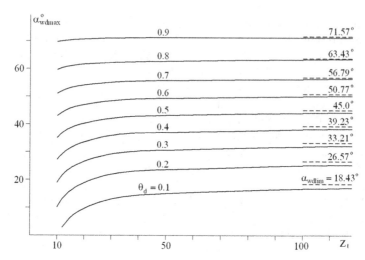

FIGURE 4.13
Maximum drive flank pressure angles α_{wdmax} and their limits α_{wdlim} when $z_t \to \infty$.

The drive and coast pressure angle maximum limits for the external spur asymmetric reversible gears with the total number of teeth $z_t \to \infty$ are as follows [28]:

$$\alpha_{wd}^{\max \lim} = \arctan K, \tag{4.9}$$

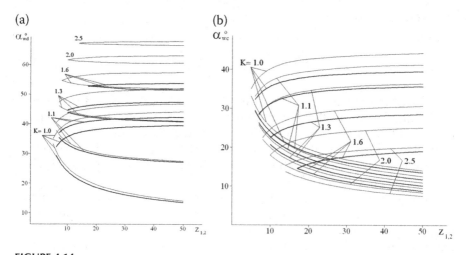

FIGURE 4.14
Minimum and maximum drive (a) and coast (b) flank pressure angles α_{wd} and α_{wc} for the external spur gears with a gear ratio $u = 1.0$, and relative tooth tip thicknesses $m_{a1,2} = 0$ (thin lines) and $m_{a1,2} = 0.25$ (thick lines).

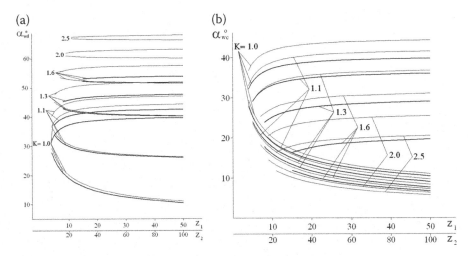

FIGURE 4.15
Minimum and maximum drive (a) and coast (b) flank pressure angles α_{wd} and α_{wc} for the external spur gears with a gear ratio $u = 2.0$, and relative tooth tip thicknesses $m_{a1,2} = 0$ (thin lines) and $m_{a1,2} = 0.25$ (thick lines).

$$\alpha_{wc}^{\max \lim} = \arctan \frac{1}{K}. \tag{4.10}$$

Figure 4.16 shows a chart of the pressure angle limits as functions of the asymmetry factor K.

Helical asymmetric tooth gears can have the transverse contact ratio $0 < \varepsilon_\alpha < 1.0$. This expands a theoretical range of the transverse pressure angle to $0° < \alpha_w < 90°$ (see Figure 4.17). A practical application of helical asymmetric tooth gears with very high transverse pressure angles (75° to 85°) is self-locking gears [32].

4.3 Contact Ratios

4.3.1 Symmetric Gearing

The theoretical minimum transverse contact ratio for involute helical gears is $\varepsilon_\alpha = 0$. In this case, the active involute profiles of the mating gear flanks are shrunk to a point, and proper tooth engagement requires helical gears with an overlap ratio $\varepsilon_\beta > 1.0$.

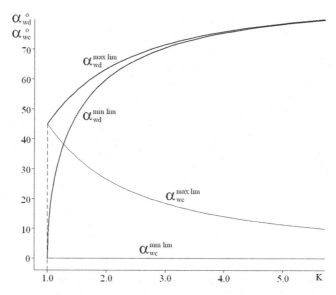

FIGURE 4.16
Drive (thick lines) and coast (thin lines) flank pressure angle limits as functions of the asymmetry factor K.

FIGURE 4.17
Helical gears with high-pressure angle. (From Kapelevich, A.L., and R.E. Kleiss, *Gear Technology*, September/October 2002, 29–35. With permission.)

In traditional gear design, the maximum transverse contact ratio is defined by the selected basic or generating rack and its X-shifts for pinion and gear. The contact ratio limit, in this case, is achieved when the numbers of teeth of the mating gears approach infinity: $z_{1,2} \rightarrow \infty$ and the mating gears become the gear racks. This limit can be defined as (see Figure 4.18)

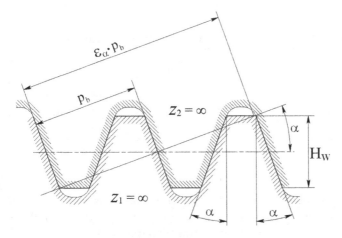

FIGURE 4.18
Contact ratio limit definition for traditionally designed gears.

$$\varepsilon_{\alpha \, \text{lim}} = \frac{H_w}{p_b \sin \alpha} \qquad (4.11)$$

or

$$\varepsilon_{\alpha \, \text{lim}} = \frac{2h_w}{\pi \sin 2\alpha}. \qquad (4.12)$$

Table 4.11 presents the contact ratio limits for different generating racks.

Direct Gear Design® expands the transverse contact ratio range. Its maximum value depends on the type of gearing (external, internal, or rack and pinion), tooth profile (symmetric or asymmetric), number of teeth, and relative tooth tip thicknesses. The highest contact ratio for gear pairs with a particular number of teeth and the relative tooth tip thicknesses is achieved at point B of the area of existence at the intersection of the interference isograms. The lower the relative

TABLE 4.11

Contact Ratio Limits for Traditionally Designed Gears

Generating rack	Pressure angle, α	Active depth coefficient, h_w	Contact ratio limit, $\varepsilon_{\alpha \text{lim}}$
Standard	14½°	2.0	2.63
Standard	20°	2.0	1.98
Standard	25°	2.0	1.66
Non-standard	28°	1.8	1.38
Non-standard	20°	2.3	2.28

TABLE 4.12

Maximum Contact Ratios for Gears with Symmetric Teeth

Number of teeth, $z_{1,2}$	5	10	15	20	30	40	50
Maximum contact ratio, ε_a	1.04	1.51	1.9	2.26	2.89	3.46	3.98
Pressure angle, $\alpha_w°$	33.14	25.31	21.72	19.53	16.85	15.21	14.05
Tooth tip angle, $\alpha_{a1,2}°$	52.56	43.4	38.55	35.35	31.21	28.53	26.59
Lowest involute angle, $\alpha_{p1,2}°$	0.0	0.0	0.0	0.0	0.0	0.0	0.0

tooth tip thicknesses $ma_{1,2}$, the higher the contact ratio at point B. It has its maximum value when the relative tooth tip thicknesses $ma_{1,2} = 0$. For external symmetric gears, in this case, the drive pitch $\theta_d = 0.5$ and the maximum transverse contact ratio is defined from Equation (3.46)

$$\arctan\left(\frac{2\pi\varepsilon_a}{z_1}\right) + u\arctan\left(\frac{2\pi\varepsilon_a}{z_2}\right) - (1+u)\arctan\left(\frac{2\pi\varepsilon_a}{z_t}\right) - \frac{\pi(2\varepsilon_a - 1)}{z_1} = 0.$$

(4.13)

Table 4.12 shows maximum transverse contact ratios for external gears with symmetric teeth (Figure 4.19).

The contact ratio for rack and pinion and internal gearings is noticeably higher than that for external gearing with the same pinion and pressure angle α_w (see Figure 4.20).

4.3.2 Asymmetric Gearing

Reversible spur asymmetric gears require the coast flank contact ratio to be $\varepsilon_{ac} \geq 1.0$. The highest contact ratio is achieved at point B of an area of

(a)   (b)   (c)   (d)

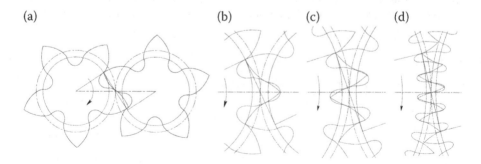

FIGURE 4.19

External symmetric gears with maximum contact ratio; (a) number of teeth $z_{1,2} = 5$, (b) $z_{1,2} = 10$, (c) $z_{1,2} = 20$, (d) $z_{1,2} = 50$.

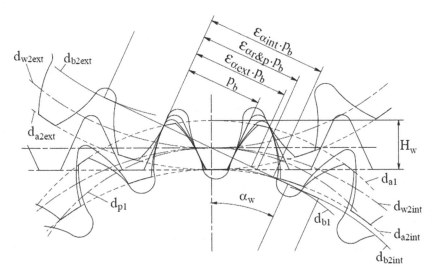

FIGURE 4.20
Contact ratio for external (index — ext), rack and pinion (r&p), and internal (int) mesh.

existence at the intersection of the drive tooth flank interference iso-grams. Below this point, coast flank interference occurs, resulting in involute profile undercut. Such undercut is permissible if the coast flank contact ratio is $\varepsilon_{ac} \geq 1.0$. The condition $\varepsilon_{ac} = 1.0$ defines the coast flank undercut profile angles $\alpha_{uc1,2}$, drive and coast flank pitch factors θ_d and θ_c, and asymmetry factor K. Figures 4.21 and 4.22 show the charts with maximum and minimum contact ratios for the spur external reversible asymmetric gears with the relative tooth tip thicknesses $m_{a1,2} = 0$ and $m_{a1,2} = 0.25$.

Table 4.13 shows maximum transverse contact ratios for spur external reversible asymmetric gears with the relative tooth tip thicknesses $m_{a1,2} = 0$. This table data indicated that the maximum drive contact ratio of the re-versible asymmetric gears is just slightly larger than it is for the symmetric gears, and asymmetry factors K of such gears is very low. Therefore, the application of asymmetric reversible gears for the drive contact ratio max-imization is not practical.

Irreversible asymmetric gears present more theoretical rather than practical interest because the benefits of their applications are not ap-parent. Table 4.14 and Figure 4.23 show the data and tooth profiles of these gears.

Figure 4.24 shows a comparison chart of maximum contact ratio and re-lated pressure angle for symmetric and asymmetric (reversible and irre-versible) gears. It indicates that irreversible asymmetric gears allow the realization of a significantly higher contact ratio.

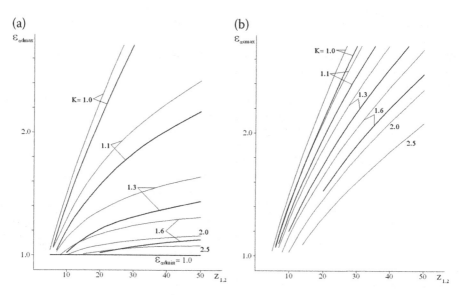

FIGURE 4.21
Minimum and maximum drive (a) and coast (b) flank contact ratios α_{wd} and α_{wc} for the external reversible asymmetric spur gears with a gear ratio $u = 1$, and relative tooth tip thicknesses $m_{a1,2}$ = 0 (thin lines) and $m_{a1,2} = 0.25$ (thick lines).

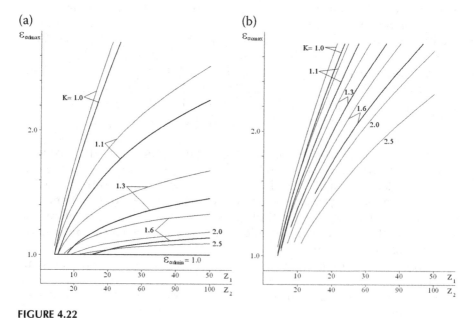

FIGURE 4.22
Minimum and maximum drive (a) and coast (b) flank contact ratios α_{wd} and α_{wc} for the external reversible asymmetric spur gears with a gear ratio $u = 2.0$, and relative tooth tip thicknesses $m_{a1,2} = 0$ (thin lines) and $m_{a1,2} = 0.25$ (thick lines).

TABLE 4.13

Maximum Drive Contact Ratios for Reversible Asymmetric Gears

Number of teeth, $z_{1,2}$	10	15	20	30	40	50
Drive contact ratio, ε_{ad}	1.53	1.931	2.288	2.924	3.49	4.015
Coast contact ratio, ε_{ac}	1.0	1.0	1.0	1.0	1.0	1.0
Drive flank pressure angle, $\alpha_{wd}°$	25.67	22.02	19.77	17.02	15.34	14.157
Coast flank pressure angle, $\alpha_{wc}°$	20.58	15.79	13.19	10.47	8.73	7.692
Drive flank tooth tip angle, $\alpha_{ad1,2}°$	43.87	38.97	35.71	31.48	28.75	26.77
Coast flank tooth tip angle, $\alpha_{ac1,2}°$	41.51	36.19	32.85	28.71	26.03	24.147
Drive flank lowest involute angle, $\alpha_{pd1,2}°$	0.0	0.0	0.0	0.0	0.0	0.0
Coast flank undercut angle, $\alpha_{uc1,2}°$	3.54	4.21	4.38	4.29	4.252	4.124
Drive flank pitch factor, θ_d	0.519	0.519	0.517	0.514	0.512	0.511
Coast flank pitch factor, θ_c	0.481	0.481	0.483	0.486	0.488	0.489
Non-contact pitch factor, θ_v	0.0	0.0	0.0	0.0	0.0	0.0
Asymmetry factor, K	1.039	1.038	1.035	1.028	1.025	1.022

TABLE 4.14

Maximum Transverse Contact Ratios of Irreversible Asymmetric Gears

Number of teeth, $z_{1,2}$	10	20	33	48
Contact ratio, ε_{ad}	2.08	3.01	4.02	5.04
Pressure angle, $\alpha_{wd}°$	33.15	25.31	20.97	18.27
Tooth tip angle, $\alpha_{ad1,2}°$	52.56	43.41	37.46	33.44
Lowest involute angle, $\alpha_{pd1,2}$	0	0	0	0

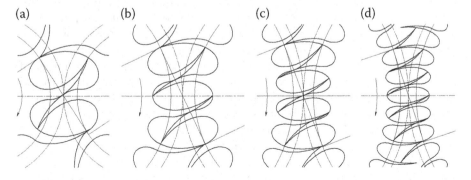

(a) (b) (c) (d)

FIGURE 4.23

Irreversible external asymmetric gears with maximum contact ratio; (a) numbers of teeth $z_{1,2} = 10$, (b) $z_{1,2} = 20$, (c) $z_{1,2} = 33$, (d) $z_{1,2} = 48$.

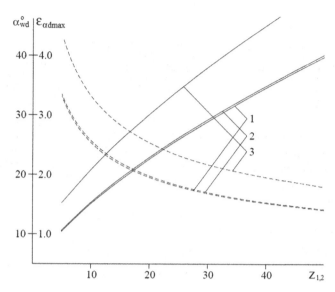

FIGURE 4.24

Maximum contact ratio $\varepsilon_{\alpha max}$ (solid curves) and related pressure angle (dashed curves) of directly designed gears; 1 – with symmetric teeth, 2 and 3 – for reversible and irreversible asymmetric gears respectively.

4.4 Practical Range of Involute Gear Parameters

Direct Gear Design® significantly expands the boundaries of involute gearing. However, in most cases, gear tooth and mesh geometry parameters do not reach their theoretical limits, because of, first, specific gear application performance requirements and, second, some material and technological constraints. For example, the application of gears with a very low number of teeth is limited by increased specific sliding velocities, resulting in low mesh efficiency, higher gear mesh temperature, and tooth flank scuffing probability. At the same time, this reduces tooth deflection under an operating load and flank impact absorption, resulting in higher noise and vibration. On the contrary, gears with a given pitch diameter and a very high number of teeth have a very small tooth size. This leads to reduced bending strength and increases gear drive assembly tolerance sensitivity when at some tolerance combination, the contact ratio can be reduced to $\varepsilon_\alpha < 1.0$, which also results in increased noise and vibration and degrades gear drive performance.

The practical maximum pressure angle and transverse contact ratio are limited by the minimum tooth tip thickness. For case hardened teeth, it is sufficient to avoid the hardening through the tooth tip. For gears made out of soft metals and plastics, it is sufficient to exclude tooth tip bending. The minimum relative tooth tip thickness typically is $m_{a1,2} = 0.25$–0.3.

4.4.1 Symmetric Gearing

A practical minimal contact ratio for conventional spur gears is about $\varepsilon_{\alpha min} = 1.1$–$1.15$. For high contact ratio (HCR) gears it is about $\varepsilon_{\alpha min} = 2.05$–$2.1$. These minimal contact ratio values are chosen to avoid its reduction below 1.0 for conventional spur gears and below 2.0 for HCR spur gears, because of manufacturing and assembly tolerances, and tooth tip chamfers or radii. These conditions also identify the practical maximum pressure angle. The practical minimal pressure angle for symmetric gears is defined by the beginning of the tooth involute

TABLE 4.15

Practical Range of α_w and ε_α for Symmetric Gears ($m_{a1,2} = 0.3$)

z_2	Parameter	z_1						
		15	20	30	40	50	70	100
15	$\alpha_{w min}$	20.33						
	$\alpha_{w max}$	33.60						
	$\varepsilon_{\alpha min}$	1.10						
	$\varepsilon_{\alpha max}$	1.77						
20	$\alpha_{w min}$	18.90	18.26					
	$\alpha_{w max}$	34.20	34.50					
	$\varepsilon_{\alpha min}$	1.10	1.10					
	$\varepsilon_{\alpha max}$	1.91	2.10					
30	$\alpha_{w min}$	15.98	17.51	21.00				
	$\alpha_{w max}$	34.80	35.01	35.58				
	$\varepsilon_{\alpha min}$	1.10	1.10	1.10				
	$\varepsilon_{\alpha max}$	2.05	2.10	2.10				
40	$\alpha_{w min}$	13.57	18.12	21.70	22.30			
	$\alpha_{w max}$	35.30	35.40	35.65	35.96			
	$\varepsilon_{\alpha min}$	1.10	1.10	1.10	1.10			
	$\varepsilon_{\alpha max}$	2.10	2.10	2.10	2.10			
50	$\alpha_{w min}$	13.90	18.5	22.30	22.70	23.10		
	$\alpha_{w max}$	35.60	35.70	35.85	36.10	36.30		
	$\varepsilon_{\alpha min}$	1.10	1.10	1.10	1.10	1.10		
	$\varepsilon_{\alpha max}$	2.10	2.10	2.10	2.10	2.10		
70	$\alpha_{w min}$	14.30	18.92	23.00	23.30	23.58	23.90	
	$\alpha_{w max}$	35.95	36.06	36.30	36.40	36.50	36.60	
	$\varepsilon_{\alpha min}$	1.10	1.10	1.10	1.10	1.10	1.10	
	$\varepsilon_{\alpha max}$	2.10	2.10	2.10	2.10	2.10	2.10	
100	$\alpha_{w min}$	14.62	19.22	23.70	23.90	24.05	24.30	24.55
	$\alpha_{w max}$	36.30	36.35	36.45	36.55	36.65	36.75	36.80
	$\varepsilon_{\alpha min}$	1.10	1.10	1.10	1.10	1.10	1.10	1.10
	$\varepsilon_{\alpha max}$	2.10	2.10	2.10	2.10	2.10	2.10	2.10

undercut, when the involute profile angles at the lowest contact points $\alpha_{pd1,2} = 0°$, where the transverse contact ratio reaches its maximum value ε_{amax}. Table 4.15 presents a practical range of pressure angles and contact ratios for spur external gears with symmetric teeth.

4.4.2 Asymmetric Gearing

The application of gears with asymmetric teeth allows the drive flank pressure angle to be increased in comparison to gears with symmetric teeth because of the coast flank pressure angle reduction. If coast flanks are not used for load transmission and may just occasionally be engaged in contact (as a result of tooth bouncing, inertial load during gear drive deceleration, etc.), the coast flank pressure angle can be as low as $\alpha_{wc} = 10°–15°$. The practical maximum drive flank pressure angles for conventional and HCR asymmetric gears are shown in Tables 4.16 and 4.17.

TABLE 4.16

Practical α_{wdmax} for Conventional Asymmetric Gears ($m_{a1,2} = 0.3$, $\varepsilon_{ad} = 1.1$, $\alpha_{wc} = 15°$)

z_2	z_1						
	15	20	30	40	50	70	100
15	43.5						
20	44.5	45.5					
30	45.9	46.4	47.3				
40	47	47.3	47.7	48.2			
50	47.6	47.8	48	48.3	48.9		
70	48	48.2	48.6	48.7	49	49.5	
100	48.7	48.9	49.2	49.5	49.1	49.6	50

TABLE 4.17

Practical α_{wdmax} for HCR Asymmetric Gears ($m_{a1,2} = 0.3$, $\varepsilon_{ad} = 2.1$, $\alpha_{wc} = 15°$)

z_2	z_1						
	20	25	30	40	50	70	100
20	19.3						
25	20.5	21.5					
30	21.5	22.4	23				
40	23	23.6	24.1	25			
50	24.1	24.6	25	25.6	26.1		
70	25.5	25.8	26.1	26.6	26.9	27.5	
100	26.7	27	27.2	27.5	27.7	28.1	28.5

(a) (b)

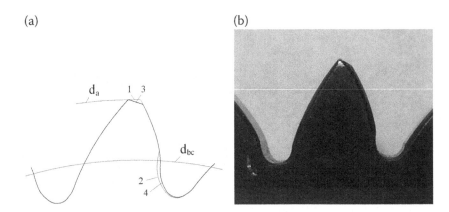

FIGURE 4.25
Asymmetric tooth with slanted tip; (a) profile, (b) photo (in this case a root fillet is not optimized); d_a – gear outer diameter, d_{bc} – coast flank base diameter; 1: circular tooth tip land, 2: slanted tooth tip land, 3: root fillet profile with undercut optimized from a trajectory of the mating tooth with the circular tip, 4: root fillet profile optimized from a trajectory of the mating tooth with the slanted tip.

The maximum drive pressure angle values in Tables 4.16 and 4.17 assume some possible small undercut of the coast flank near the root, especially for gears with a low number of teeth (15–30). However, this does not reduce the coast flank contact below $\varepsilon_{ac} = 1.0$. This undercut can be reduced or completely eliminated by using slanted tooth tips (Figure 4.25) [27]. This increases the tooth tip land and reduces bending stress. These slanted tooth tips can be produced by the special topping gear cutter (hob) or by the secondary (after the tooth hobbing) tooth tip milling operation.

If the coast flanks are used in normal operating load transmission, as in, for example, idler or planet gears, the asymmetry factor K and practical range of pressure angles and contact ratios are defined based on specific application requirements (see Section 5.1.1).

5

Gear Geometry Optimization

Direct Gear Design® is not limited by any preselected tooling parameters or fabrication process requirements. Its tooth geometry boundaries are considerably expanded in comparison to the traditional standard gear design methods. It allows definition of an optimal tooth shape for a specific custom gear application. Such gears require special tooling and typically more expensive in production. This potential increase in manufacturing cost must be justified by significant performance improvement over the best gears designed by standards. All these reasons make the gear geometry optimization, described in this chapter, an essential part of Direct Gear Design®.

A starting point of gear tooth geometry optimization is to establish a set of priorities for a specific gear drive application. This "wish list" may include, for example:

- gear transmission density maximization that requires minimizing gear drive size and weight for a given transmitted load and gear ratio, or maximizing the load for a given gear drive size;
- accommodation into a required space or envelope (typically for the integrated gear drives);
- noise and vibration reduction;
- cost reduction;
- increased life and reliability;
- other performance enhancement requirements.

These application priorities dictate the tooth geometry optimization techniques. As a part of a gear drive design, the tooth geometry optimization techniques should be done in combination with other gear transmission component optimization. This chapter describes the macro and microgeometry optimization of asymmetric tooth gears.

5.1 Tooth Macrogeometry

The gear tooth macrogeometry optimization includes:

- definition of optimal parameters of the involute flanks for the gear pair size minimization;
- selection of an asymmetry factor for balancing the performance of the drive and cost flanks of asymmetric gears;
- determination of the drive contact ratio for optimal tooth pair load distribution and transmission error variation minimization;
- tooth root fillet profile optimization for stress minimization;
- root stress balance.

5.1.1 Involute Flank Profile

Involute tooth flank profile parameters influence many gear drive performance characteristics. They affect the contact surface endurance (pitting and scuffing resistance), tooth root bending fatigue resistance, profile sliding and gear efficiency, tooth flexibility and load sharing, and vibrations and noise.

5.1.1.1 Gear Pair Size Reduction

Gear pair size reduction is typically limited by the tooth flank surface endurance defined by pitting and scuffing. Both types of tooth surface defects depend on the contact stress and sliding velocity. Application of a higher operating pressure angle ($a_w = 25$–$30°$ for gears with symmetric teeth and $a_{wd} = 30$–$45°$ for drive flank of gears with asymmetric teeth) leads to Hertz contact stress reduction. It allows an increase of the transmitted torque and reducing of the gear size (diameter, face width, or both), maintaining an acceptable contact stress level for the given operating conditions and gear materials. It also makes gear teeth more stubby, with the reduced height and increased thickness at the root area, reducing the root bending stress. High pressure angle gear pairs have a relatively low contact ratio of $\varepsilon_a = 1.1$–1.3 and profile sliding that decreases scuffing probability. Drawbacks of the application of high pressure angle gears are large separating load taken by supporting bearings and high gear mesh stiffness that reduces tooth engagement impact dampening and leads to higher noise and vibration.

Another way of minimizing gear pair size is the application of gears with a high contact ratio (HCR). Conventional spur involute gears have a transverse contact ratio $1.0 < \varepsilon_a < 2.0$ when one or two pairs of teeth are in simultaneous contact. The HCR spur gears have a transverse contact ratio $\varepsilon_a > 2.0$ (typically around 2.05–2.2) when two or three pairs of teeth are in simultaneous contact. The HCR gears provide load sharing between gear pairs in contact, and according to [8], a maximum gear pair load does not exceed about 2/3 of the total transmitted load. The HCR gears have a relatively low operating pressure angle ($a_w = 18$–$23°$ for gears with symmetric teeth and $a_{wd} = 23$–$28°$ for drive flank of gears with asymmetric teeth), an increased tooth height, and a reduced thickness at the tooth root. As a result, they have increased tooth deflection, providing a better load sharing between engaged tooth pairs and allowing reduction of contact and bending stresses, and also a noise and vibration reduction that makes them applicable for aerospace gear transmissions [8,38]. However, this requires the HCR gears to be sufficiently accurate, having the base pitch variation lower than the tooth deflection under the required operating load to provide load sharing between the contacting tooth pairs. The HCR gears also have some limitations. The minimum number of teeth should be at 20 in order to provide a transverse contact ratio $\varepsilon_\alpha \geq 2.0$. A long tooth addendum and low operating pressure angle result in higher sliding velocity that increases the scuffing probability and mesh power losses. However, according to [39], "despite their higher sliding velocities high contact ratio gears can be designed to levels of efficiency comparable to those of conventional gears while retaining their advantages through proper selection of gear geometry."

Attempts to use the buttress asymmetric HRC gears in the sun/planet mesh of the planetary gear stage for noise and vibration reduction were not successful [40,41]. This could be explained by the high stiffness of buttress teeth that have a low drive pressure angle and a high coast pressure angle. It seems that a more rational approach for many applications of asymmetric gears is to have the drive tooth flanks with a higher pressure angle than the coast ones. In the case of epicyclic gear stages where a planet gear has both driving flanks, a higher pressure angle should be used in the sun/planet gear mesh, and a lower pressure angle in the planet/ring gear mesh. Application of this approach to the asymmetric HCR gears allows designing of the coast tooth flanks independently to reduce gear tooth stiffness for a better tooth load sharing. Table 5.1 shows the gear tooth profile geometry of the high pressure angle and high contact ratio gears.

5.1.1.2 Asymmetry Factor Selection

Asymmetric tooth profiles make it possible to increase the operating pressure angle beyond the conventional symmetric gear limits. This reduces drive flank contact stress and sliding velocity, increasing the tooth surface endurance to pitting and scuffing, and providing the maximized gear transmission density.

The selection of the asymmetry factor K depends on the gear pair operating cycle that is defined by RPM and transmitted load in the main, and reversed directions, and gear drive life requirements [42]. If the gear tooth is equally loaded in both the main and reversed rotation directions, asymmetric tooth profiles should not be considered. Table 5.2 illustrates the bidirectional and unidirectional load transmission cases [43].

TABLE 5.1

Geometry Parameters of High Pressure Angle and High Contact Ratio Gears

Type	High pressure angle		High contact ratio (HCR)	
Tooth form	**Symmetric**	**Asymmetric**	**Symmetric**	**Asymmetric**
Tooth profile				
Number of teeth pinion and gear $z_{1,2}$	35	35	35	35
Pressure angle α_{wd}	30°	42°	22°	26°
α_{wc}	30°	14°	22°	12°
Asymmetry factor K	1.0	1.3	1.0	1.09
Transverse contact ratio ε_{ad}	1.39	1.25	2.04	2.02
ε_{ac}	1.39	2.02	2.04	2.16

TABLE 5.2

Gears for Bidirectional and Unidirectional Load Transmission

Case #		1	2	3	4	5
Load transmission		Bidirectional		Mostly unidirectional	Unidirectional	
Loaded flanks		Both	Both	Drive, lower coast load	Drive, very low coast load	Drive flank only
Tooth profile		Symmetric (baseline)	Symmetric	Asymmetric	Asymmetric	Asymmetric
Gear mesh		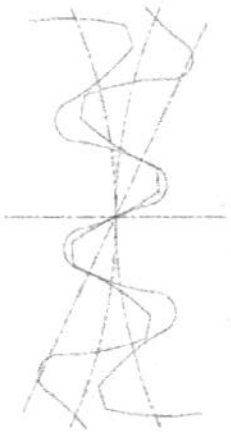	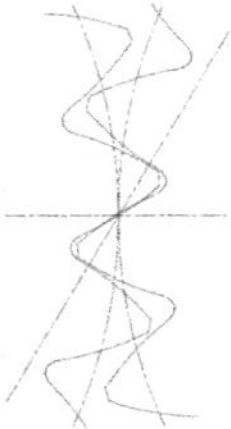	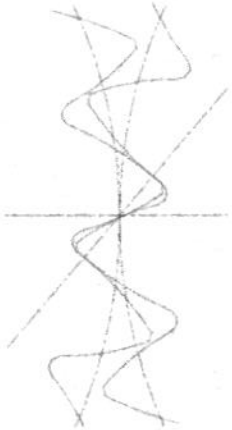	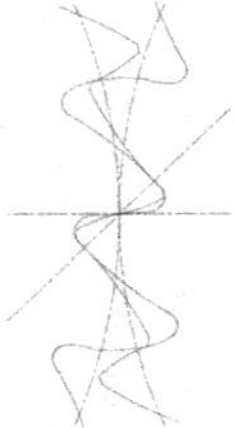	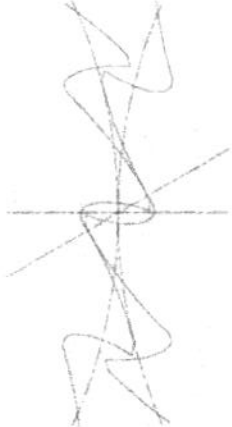
Pressure angle	α_{wd}	25°	32°	40°	46°	60°
	α_{wc}	25°	32°	24°	10°	*
Asymmetry coefficient, K		1.0	1.0	1.19	1.42	*
Transverse contact ratio	ε_{ad}	1.35	1.2	1.2	1.2	1.2
	ε_{ac}	1.35	1.2	1.44	1.0	*
Hertz contact stress, %	Drive	100	92	88	86	94
	Coast	100	92	102	150	*
Bearing load, %	Drive	100	107	118	130	181
	Coast	100	107	99	92	*
Specific sliding velocity, %	Drive	100	94	75	68	49
	Coast	100	94	108	97	*

Note:

* Coast flank mesh does not exist

Cases 1 and 2

The gear teeth are symmetric, and their surface durability is identical for both drive and coast flanks. Case #1 presents the traditionally designed 25° pressure angle gears with the full radius fillet. This case is considered as a baseline, and its Hertz contact stress, bearing load, and specific sliding velocity are assumed as 100% for comparison with other gear pairs. This type of gear profile is used in the aerospace industry because it provides better bending strength and flank surface endurance in comparison with the standard 20° pressure angle gears typical for commercial applications. Case #2 is the high 32° pressure angle symmetric gears, optimized by the Direct Gear Design® method. Its Hertzian contact stress is about 8% lower, and the specific sliding velocity is about 6% lower than for the baseline gear pair. It should provide better flank tooth surface pitting or scoring resistance. However, the bearing load is 7% higher.

Case 3

These asymmetric gears are for mostly unidirectional load transmission with a 40° operating pressure angle driving tooth flanks providing 12% contact stress and 25% sliding velocity reduction. At the same time, the contact stress and sliding velocity of the coast flanks are close to these parameters of the baseline gears and should provide tooth surface load capacity similar to that for the baseline gears. This type of gear may find applications for drives with one main load transmission direction, but is also capable of carrying a lighter load for shorter periods in the opposite load transmission direction.

Case 4

These asymmetric gears have a 46° operating drive pressure angle that allows the reduction of the contact stress by 14% and sliding velocity by 32%. The disadvantage of such gear teeth is a high (+30%) bearing load. These types of gears are only for unidirectional load transmission. Their 10° coast pressure angle flanks have low load capacity. They may find applications for drives with only one load transmission direction that may occasionally transmit a very low load by the coast flanks tooth contact.

Case 5

These asymmetric gears have only drive tooth flanks with an extreme 60° operating pressure angle and no involute coast tooth flanks at all. As a result, the bearing load is significant.

There are many applications, as described in Case 2, where a gear pair transmits load in both directions, but with significantly different load magnitude and duration (Figure 5.1). In this case, the asymmetry factor K is defined by equalizing accumulated tooth surface damage defined by operating contact stress and the number of tooth flank load cycles in each load

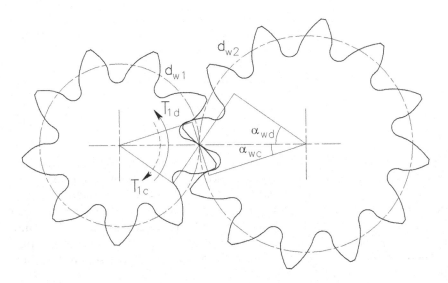

FIGURE 5.1
Asymmetric gear pair; T_{1d} and T_{1c}: pinion torques applied to the drive and coast tooth flanks. (From Kapelevich, A.L., Asymmetric gears: parameter selection approach. *Gear Technology*, June/July 2012, 48–51. With permission.)

transmission direction. In other words, a contact stress safety factor S_H should be the same for the drive and coast tooth flanks. This condition can be presented as follows:

$$S_H = \frac{\sigma_{HPd}}{\sigma_{Hd}} = \frac{\sigma_{HPc}}{\sigma_{Hc}}, \tag{5.1}$$

where σ_{Hd} and σ_{Hc} are the operating contact stresses for the drive and coast tooth flanks, and σ_{HPd} and σ_{HPc} are the permissible contact stresses for the drive and coast tooth flanks that depend on a number of load cycles.

Then from (5.1)

$$\frac{\sigma_{Hd}}{\sigma_{Hc}} = \frac{\sigma_{HPd}}{\sigma_{HPc}}. \tag{5.2}$$

The contact stress at the pitch point [44] is

$$\sigma_H = z_H z_E z_\epsilon z_\beta \sqrt{\frac{F_t}{d_{w1} b_w} \frac{u \pm 1}{u}}, \tag{5.3}$$

where

$$z_H = \sqrt{\frac{2 \cos \beta_b \cos \alpha_{wt}}{\cos^2 \alpha_t \sin \alpha_{wt}}} \tag{5.4}$$

is the zone factor that for the directly designed spur gears can be defined as

$$z_H = \frac{2}{\sqrt{\sin 2\alpha_w}}, \tag{5.5}$$

Z_E is the elasticity factor that takes into account gear material properties (Modulus of elasticity and Poisson ratio); Z_ε is the contact ratio factor, its conservative value for spur gears is $Z_\varepsilon = 1.0$; Z_β is the helix factor, for spur gears $Z_\beta = 1.0$; F_t is the nominal tangent load, which at the pitch diameter d_{w1} is $F_t = 2T_1/d_{w1}$; T_1 is the pinion torque; b_w is the contact face width; and the "+" sign is for external gearing and the "−" sign for external gearing.

Then for the directly designed spur gears, the contact stress at the pitch point can be presented as follows:

$$\sigma_H = z_E \frac{2}{d_{w1}} \sqrt{\frac{2T_1}{b_w \sin 2\alpha_w} \frac{u \pm 1}{u}}. \tag{5.6}$$

Some parameters of this equation, Z_E, d_{w1}, b_w, and u, do not depend on the load transmission direction, and Equation (5.2) for the pitch point contact can be presented as follows:

$$\frac{\sin 2\alpha_{wc}}{\sin 2\alpha_{wd}} = A, \tag{5.7}$$

where the parameter A is

$$A = \frac{T_{1c}}{T_{1d}} \left(\frac{\sigma_{HPd}}{\sigma_{HPc}} \right)^2. \tag{5.8}$$

According to the ISO 6336–2006 standard [44], "the permissible stress at limited service life or the safety factor in the limited life stress range is determined using life factor Z_{NT}." This allows replacement of the permissible contact stresses in Equation (5.8) for the life factors

$$A = \frac{T_{1c}}{T_{1d}} \left(\frac{Z_{NTd}}{Z_{NTc}} \right)^2. \tag{5.9}$$

When parameter A is defined, and the drive pressure angle is selected, the coast pressure angle is calculated by Equation (5.7) and the asymmetry coefficient K from a combined solution of (5.7) and (2.104):

$$K = \frac{\sqrt{1 + \sqrt{1 - A^2 \sin^2 2\alpha_{wd}}}}{\sqrt{2} \cos \alpha_{wd}}. \tag{5.10}$$

Example 1:

The drive pinion torque T_{1d} is two times greater than the coast pinion torque T_{1c}. The drive tooth flank has 10^9 load cycles, and the coast tooth flank has 10^6 load cycles during the life of the gear drive. From the S–N curve [44] for steel gears, an approximate ratio of the life factors $Z_{NTd}/Z_{NTc} = 0.85$. Then the coefficient $A = 0.85^2/2 = 0.36$. Assuming the drive pressure angle is $\alpha_{wd} = 36°$, the coast pressure angle from Equation (5.6) is $\alpha_{wc} = 10°$ and the asymmetry factor from Equation (5.9) is $K = 1.22$.

If, as a result of the application of this asymmetric tooth flank optimization technique, a coast pressure angle $<<10°$, its selection should be driven by some other criteria, for instance, by suitability for gear fabrication.

In some mostly irreversible gear drives (Case 3 in Table 5.2), the coast tooth flanks are loaded by the system inertia during the gear drive deceleration or the tooth bouncing in the high RPM drives. This coast tooth flank load can be significant and should be taken into consideration while defining the asymmetry factor K.

If the gear drive is irreversible and the coast tooth flanks never transmit any load (Case 4 in Table 5.2), the asymmetry factor is defined only by the drive flank geometry. In this case, an increase of the drive flank pressure could be limited by a minimum selected contact ratio and a separating load applied to the bearings. The application of a very high drive flank pressure angle results in the reduced coast flank pressure angle and possibly its involute profile undercut near the tooth root. Another limitation of the asymmetry factor K of the irreversible gear drive is growing compressive bending stress at the coast flank root. Usually, for conventional symmetric gears, compressive bending stress does not present a problem because its allowable limit is significantly higher than that for the tensile bending stress. However, for asymmetric tooth gears, it may become an issue, especially for gears with thin rims.

In the unidirectional chain gear drive (Figure 5.2), the idler gear transmits the same load by both tooth flanks. This arrangement seems unsuitable for asymmetric gear applications. However, in many cases, the idler's mating input pinion and output gear have significantly different numbers of teeth. This allows equalizing contact stresses on opposite flanks of the idler gear asymmetric teeth to achieve maximum load capacity.

Equation (5.6) is used to define the pitch point contact stress in the pinion/idler gear mesh,

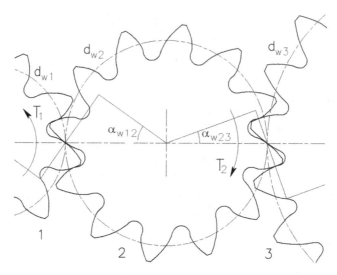

FIGURE 5.2
Chain gear attangement; 1: input pinion; 2: idler gear; 3: output gear; T_1: pinion torque; T_2: idler gear torque. (From Kapelevich, A.L., Asymmetric gears: parameter selection approach. *Gear Technology*, June/July 2012, 48–51. With permission.)

$$\sigma_{H12} = z_E \frac{2}{d_{w1}} \sqrt{\frac{2T_1}{b_{w12} \sin 2\alpha_{w12}} \frac{u_{12} + 1}{u_{12}}}, \qquad (5.11)$$

and in the idler/output gear mesh,

$$\sigma_{H23} = z_E \frac{2}{d_{w2}} \sqrt{\frac{2T_2}{b_{w23} \sin 2\alpha_{w23}} \frac{u_{23} + 1}{u_{23}}}, \qquad (5.12)$$

or, ignoring gear mesh losses,

$$\sigma_{H23} = z_E \frac{2}{u_{12}d_{w1}} \sqrt{\frac{2u_{12}T_1}{b_{w23} \sin 2\alpha_{w23}} \frac{u_{23} + 1}{u_{23}}}, \qquad (5.13)$$

where b_{w12} and b_{w23} are the contact face widths in the pinion/idler gear and the idler/output gear meshes, accordingly; $u_{12} = z_2/z_1$ is a gear ratio in the pinion/idler gear mesh; $u_{23} = z_3/z_2$ is a gear ratio in the idler/output gear mesh; and z_1, z_2, and z_3 are numbers of teeth of the input, idler, and output gears.

Numbers of the idler gear tooth load cycles and permissible contact stresses, in this case, are equal in both meshes, and Equation (5.2) can be presented as $\sigma_{H12} = \sigma_{H23}$. Then assuming that all gears are made from the same material, the idler gear pressure angle ratio is defined by

$$\frac{\sin 2\alpha_{w23}}{\sin 2\alpha_{w12}} = B \tag{5.14}$$

where

$$B = \frac{b_{w12}}{b_{w23}} \frac{u_{23} + 1}{u_{23}(u_{12} + 1)} \tag{5.15}$$

is a parameter that reflects the gear ratios u_{12} and u_{23}, and contact face widths b_{w12} and b_{w23} in the pinion/idler gear and idler/output gear meshes, respectively.

Then considering Equation (2.104) the asymmetry factor K can be presented as follows:

$$K = \frac{\sqrt{1 + \sqrt{1 - B^2 \sin^2 2\alpha_{wd}}}}{\sqrt{2} \, \cos \alpha_{wd}}. \tag{5.16}$$

If $z_1 = z_3$ and $b_{w12} = b_{w23}$, then both the coefficient B and the asymmetry factor K are equal to 1.0, and gear teeth are symmetric. If $z_1 \neq z_3$ or $b_{w12} \neq b_{w23}$, the application of asymmetric gears can be considered.

Example 2:

The pinion number of teeth is $n_1 = 9$, the idler gear number of teeth is $n_2 = 12$, the output gear number of teeth is $n_3 = 20$, the contact face width ratio is $b_{w12}/b_{w23} = 1.2$. This makes parameter $B = 0.82$. Then assuming the pinion/idler gear mesh pressure angle is $\alpha_{w12} = 35°$, the idler/output pressure angle from (5.14) is $\alpha_{w23} = 25.32°$, and the asymmetry factor from Equation (5.16) is $K = 1.10$.

This contact stress equalization technique can be also applied for the unidirectional epicyclic gear stage (Figure 5.3). In this case, the planet gear works like the idler gear engaged with the sun gear and ring gear. The nature of a contact in the sun/planet gear mesh is different from that in the planet/ring gear mesh. Convex flanks of the external tooth planet gear are in simultaneous contact with convex flanks of the external tooth sun gear and concave flanks of the internal tooth ring gear. This results in a significant difference in Hertzian contact stresses for symmetric tooth gears. However, the application of asymmetric tooth gears allows equalizing contact stresses on opposite flanks of the planet gear teeth to achieve maximum load capacity.

In this case, the asymmetry factor K is also defined by Equation (5.16), where parameter B is

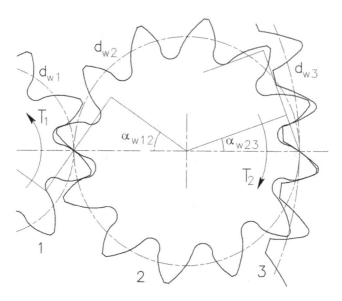

FIGURE 5.3
Planetary gear attangement; 1: sun gear; 2: planet gear; 3: ring gear; T_1: sun gear torque; T_2: planet gear torque. (From Kapelevich, A.L., Asymmetric gears: parameter selection approach. *Gear Technology*, June/July 2012, 48–51. With permission.)

$$B = \frac{b_{w12}}{b_{w23}} \frac{u_{23} - 1}{u_{23}(u_{12} + 1)}, \tag{5.17}$$

where b_{w12} and b_{w23} are the contact face widths of the sun/planet gear and planet/ring gear meshes; $u_{12} = z_2/z_1$ is the gear ratio in the sun/planet gear mesh; $u_{23} = z_3/z_2$ is the gear ratio in the planet/ring gear mesh; and $z_1, z_2,$ and z_3 are numbers of teeth of the sun, planet, and ring gears.

In a typical epicyclic gear stage $z_2 = (z_3 - z_1)/2$. This allows simplification of Equation (5.17):

$$B = \frac{b_{w12}}{u_{13}b_{w23}}, \tag{5.18}$$

where $u_{13} = z_3/z_1$ is the epicyclic stage gear ratio. The ring gear number of teeth z_3 is much greater than the sun gear number of teeth z_1. This makes an epicyclic gear stage suitable for taking advantage of asymmetric tooth gear application.

Example 3:

The sun gear number of teeth is $n_1 = 9$, the planet gear number of teeth is $n_2 = 12$, the ring gear number of teeth is $n_3 = 33$, and the contact face width ratio is $b_{w12}/b_{w23} = 1.8$. This makes parameter $B = 0.49$. Then assuming the sun/planet gear mesh pressure angle is $a_{w12} = 40°$, the planet/ring gear mesh pressure angle from (5.14) is 14.5°, and the asymmetry factor from Equation (5.16) is $K = 1.26$.

While utilizing this asymmetric tooth flank optimization technique, it is important to know that selecting a relatively low sun/planet gear mesh pressure angle may result in the unacceptable low planet/ring gear mesh pressure angle (<<10°). In such cases, selection of the planet/ring gear mesh pressure angle should be driven by other criteria, for example, by suitability for gear fabrication.

5.1.1.3 Gear Mesh Efficiency Maximization

Gear mesh power losses depend on the gear tooth geometry and friction coefficient. For spur gears, the percent of mesh losses for the drive flanks is defined in Equation (3.25). In the optimized asymmetric gear pair, maximum mesh efficiency (minimum of mesh losses) is achieved if the drive flank specific sliding velocities at the start of the approach action and the end of the recess action are equalized: $H = H_{sd} = H_{td}$. Then Equation (3.25) can be presented as

$$P_{td} = \frac{50fH}{\cos \alpha_{wd}}. \tag{5.19}$$

If this equation is solved with Equations (2.130), (2.132), (3.27), and (3.28), the mesh power loss percent for the drive flanks of spur gears can be defined as

$$P_{td} = 50f\pi\varepsilon_{\alpha d}\left(\frac{1}{z_1} \pm \frac{1}{z_2}\right), \tag{5.20}$$

and the maximized gear mesh efficiency as follows:

$$E_{d\max} = 100 - P_{td} = 100 - 50f\pi\varepsilon_{\alpha d}\left(\frac{1}{z_1} \pm \frac{1}{z_2}\right), \tag{5.21}$$

where signs ± are for external gears and internal gears, respectively.

Equation (5.21) indicates that for spur gears, the optimized maximum efficiency gear geometry depends only on the type of gearing (external or internal), numbers of teeth, and drive flank transverse contact ratio $\varepsilon_{\alpha d}$. For helical gears, the mesh power losses are described by Equation (3.26). For the gear pair optimized for maximum efficiency, these power losses of the drive flanks can be presented as

$$P_{td} = 50f\pi\varepsilon_{\alpha d}\cos^2\beta_w\left(\frac{1}{z_1} \pm \frac{1}{z_2}\right),\tag{5.22}$$

where signs ± are for external gears and internal gears, respectively.

For symmetric tooth gears, Equations 5.19 – 5.22 should be used without the subscripted index "*d*".

The maximum gear mesh efficiency of the spur gears is limited by the minimum transverse contact ratio $\varepsilon_\alpha = 1.0$. Helical gears make it possible to increase the gear mesh efficiency by reducing $\varepsilon_\alpha < 1.0$ and increasing the operating helix angle β_w. Such helical gears can reduce automotive transmission power losses [45].

5.1.1.4 Contact Ratio Optimization

The contact ratio of spur gears is an essential parameter that affects the gear drive performance. If influences the load capacity, efficiency, and noise and vibration. The low and high contact ratio spur gears are described in Sections 4.4 and 5.1.1.1, where a nominal contact ratio is considered as designed without the influence of tooth deflections under the operating load. It is defined by Equations (2.85–2.87) for symmetric gears and Equations (2.124–2.129) for asymmetric gears. The goal is to find an optimal contact ratio solution that would provide high load capacity and efficiency, as well as low transmission error variation, which is the main cause of gear noise and vibration.

This section introduces and analyzes an effective contact ratio that is defined considering bending and contact gear tooth deflections [46]. The effective contact ratio describes the actual duration of a mating tooth pair contact. It can be defined as the ratio of the tooth engagement angle to the angular pitch. The tooth engagement angle is a gear rotation angle from the start of the gear tooth engagement with the mating gear tooth to the end of the tooth engagement. The effective contact ratio is (see Figure 5.4):

$$\varepsilon_{\alpha de} = \frac{\varphi_1}{360°/z_1} = \frac{\varphi_2}{360°/z_2},\tag{5.23}$$

where φ_1 and φ_2 – pinion and gear tooth engagement angles,
$360°/z_1$ and $360°/z_2$ – pinion and gear angular pitches.

If the transmitted load is low and not sufficient to produce pinion and gear tooth deflections (Figure 5.4a), the starting point of tooth contact is located at point 1 of the active contact line. At this point, the gear tooth tip at diameter d_{a2} touches the pinion tooth flank at the lowest contact point diameter d_{ld1}. The tooth engagement continues up to the end of the active contact line at the point 2, where the pinion tooth tip at diameter d_{a1} meets the gear tooth flank at the lowest contact point diameter d_{ld2}. In this case, the

(a)

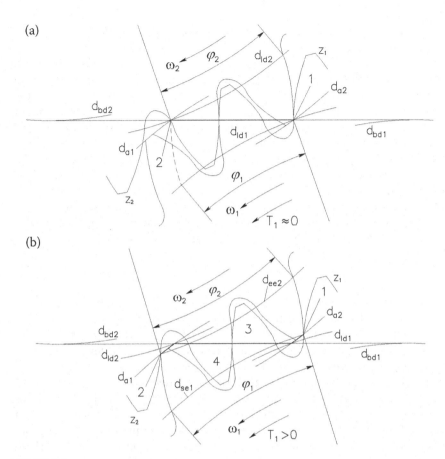

(b)

FIGURE 5.4
Pinion and gear tooth engagement angles φ_1 and φ_2; (a) unloaded or lightly loaded tooth contact; (b) loaded tooth contact; ω_1 and ω_2 – pinion and gear rotation directions; T_1: pinion torque; thick line: path of contact point from start of tooth engagement 1 to end of tooth engagement 2; dashed line: pinion tooth drive flank at the end of tooth engagement; 3 and 4: loaded/deflected teeth of pinion and gear; d_{ld1} and d_{ld2}: pinion and gear lowest contact point diameters; d_{se1}: start engagement diameter; d_{ee2}: end engagement diameter.

effective contact ratio is equal to the nominal contact ratio and Equation (5.23) can be converted into Equations (2.91) or (2.92) for symmetric gears and into Equations (2.136) or (2.137) for asymmetric gears.

If the substantial transmitted load results in pinion and gear tooth deflections (teeth 3 and 4 in Figure 5.4b), the gear tooth tip at diameter da_2 touches the pinion tooth flank at point 1 that lays at the start engagement diameter $dse1$ higher than the lowest contact point diameter d_{ld1}. Then the tooth pair contact point slides down along the pinion drive flank to its diameter d_{ld1}, making its path on the arc of the gear tip tooth diameter da_2. The tooth pair contact point reaches the nominal contact line that is tangent to both base diameters d_{bd1} and d_{bd2} and starts moving along it until the contact point coincides with the pinion tooth tip diameter d_{a1} and the mating point at the gear lowest contact point diameter d_{ld2}. Then the tooth pair contact point is moving up along the gear drive flank to the end of the tooth pair engagement point 2, making its path on the arc of the pinion tip tooth diameter d_{a1}. The loaded gear tooth pair has a longer contact path than the unloaded one by two additional portions at the start and end of the tooth engagement that lay outside of the nominal contact line. This extension of the contact path is a result of the deflection of the loaded teeth, and it increases the effective contact ratio compared to the nominal contact ratio.

Transmission error is the angular difference between the actual position of the driven gear and its ideal position (if the gear pair is perfectly conjugated), projected on the line of contact and defined as [35]

$$TE = r_{bd2}(\theta_2 - u\theta_1),\qquad(5.24)$$

where θ_1 and θ_2 are the driving pinion and driven gear rotation angles, r_{bd2} is the drive tooth flank base radius of the driven gear.

A typical spur gear transmission error chart is shown in Figure 5.5.

The effective contact ratio and transmission error are influenced by manufacturing tolerances and operating conditions, including deflections of gears (teeth and body) and other gearbox components under operating load, dynamic loads and inertia, temperature, etc. Dr. Yuriy V. Shekhtman has developed a calculation procedure for the definition of the effective contact ratio and transmission error, considering only the bending and contact tooth deflections. According to his approach, each angular position of the driven

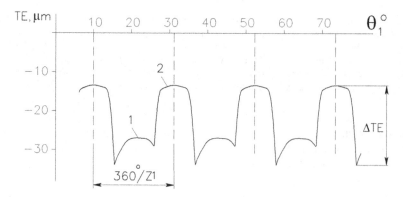

FIGURE 5.5
Typical spur gear pair transmission error (TE) chart; ΔTE: transmission error variation; 1: single tooth pair contact, 2: double tooth pair contact. (Kapelevich A.L. and Y.V. Shekhtman. Analysis and optimization of contact ratio of asymmetric gears, *Gear Technology*. March/April 2017, 66–71. With permission.)

gear relative to the driving gear is iteratively defined by equalizing a sum of the tooth contact load moments of each gear to its applied torque. The related tooth contact loads are also iteratively defined to conform to tooth bending and contact deflections. The tooth bending deflection in each contact point is determined based on the FEA calculated flexibility, and the tooth contact deflection is calculated by the Hertz equation.

The Direct Gear Design® method allows considering the drive flank nominal contact ratio as one of the gear design input parameters. Maximum gear mesh efficiency is achieved when the specific sliding velocities are equalized and defined by Equation (5.21). To define the optimal contact ratio the comparable gear sets are assumed to have identical maximized mesh efficiency E, average friction coefficient f, and gear ratio u. Then Equation (5.21) for the external spur asymmetric gears can be converted to

$$\frac{\varepsilon_{ad}}{z_1} = \left(1 - \frac{E}{100}\right)\frac{2u}{f\pi(1+u)} = const. \tag{5.25}$$

This criterion is used to analyze the parameters of external spur gear sets with asymmetric teeth. Comparable asymmetric tooth gear sets have different numbers of teeth and identical center distance a_w, gear ratio u, coast flank pressure angle α_{wc}, minimal required relative tooth tip thicknesses $m_{a1,2}$, average friction coefficient f, and gear mesh efficiency E, pinion and gear material properties, and equalized specific sliding velocities H_{sd} and H_{td}. The face widths b_1 and b_2 are defined to approximately equalize the pinion and gear tooth bending stresses considering the optimized root fillets (see Section 5.1.2). Since the center distance a_w is identical for all considered gear sets, the operating modules are inversely proportional to a number of the pinion teeth and defined from Equation (2.53) as follows:

$$m = \frac{2a_w}{z_1(1+u)}. \tag{5.26}$$

The operating pitch diameter tooth thickness ratio is

$$TTR = S_{w1}/p_w = S_{w1}/(S_{w1} + S_{w2}), \tag{5.27}$$

where S_{w1} and S_{w2} are the pinion and gear tooth thicknesses at the operating pitch diameters, p_w is the operating circular pitch.

The optimized TTR value is defined to simultaneously equalize the specific sliding velocities H_{sd} and H_{td}, and the pinion and gear relative tooth tip thicknesses $m_{a1,2}$. The maximized drive flank pressure angle α_{wd} is defined to minimize the contact stress. It must also provide a nominal drive contact ratio ε_{ad} defined by Equation (2.124), and the preselected values of the coast flank pressure angle α_{wc}, and pinion and gear relative tooth tip thicknesses $m_{a1,2}$.

The bearing load is

$$F = 2T_1/d_{bd1},$$ (5.28)

where T_1 is the pinion operating torque, d_{bd1} is the pinion drive flank base diameter.

Load sharing factor is

$$L = F_{cmax}/F,$$ (5.29)

where F_{cmax} is the maximum normal load in the single tooth pair contact.

If the drive flank effective contact ratio $\varepsilon_{ade} < 2.0$, the load sharing factor $L = 1.0$.

Table 5.3 presents gear geometry and load parameters for 15 gear pairs that are defined to satisfy preselected comparison conditions, listed in the top field of this table. Gear pairs are divided into four groups based on a contact ratio type: low, medium, transitional, and high contact ratio. Four gear sets, one from each group, are selected to define the tooth contact and bending stresses, and transmission error variation under variable operating loads to find a gear set with an optimal contact ratio.

Parameters of the selected gear sets are highlighted in the bold font in Table 5.3:

- Gear set 1 with a low drive contact ratio ($\varepsilon_{ad} = 1.19$ and $\varepsilon_{ade} = 1.33$) has a 15-tooth pinion and 30-tooth gear.
- Gear set 2 with a medium drive contact ratio ($\varepsilon_{ad} = 1.51$ and $\varepsilon_{ade} = 1.69$) has a 19-tooth pinion and 38-tooth gear.
- Gear set 3 with a transitional drive contact ratio ($\varepsilon_{ad} = 1.83$ and $\varepsilon_{ade} = 2.04$) has a 23-tooth pinion and 46-tooth gear. It is called transitional because it has a nominal drive contact ratio < 2.0. Such gears under a low load have one or two mating tooth pairs in contact. When a load is increased to its operating level and tooth deflections are increased respectively, the transitional contact ratio gears have two or three mating tooth pairs simultaneously engaged in contact. This results in tooth load sharing and a single tooth pair contact load reduction.
- Gear set 4 with a high drive contact ratio ($\varepsilon_{ad} = 2.15$ and $\varepsilon_{ade} = 2.40$) has a 27-tooth pinion and 54-tooth gear.

Figure 5.6 shows the selected gear set meshes in the same scale.

The main gear pair parameters vs. pinion number of teeth charts are shown in Figures 5.7–5.12. With increasing numbers of the pinion teeth, the

TABLE 5.3

Comparable Asymmetric Gear Set Parameters

Preselected gear set parameters and conditions Center distance: 150 mm; Gear ratio: 2:1; Coast pressure angle: 15°; Pinion and gear face widths: 35 mm and 30 mm; Relative tooth tip thickness of pinion and gear: 0.30; Average friction coefficient: 0.05; Gear mesh efficiency: 99%; Pinion and gear material properties: Modulus of elasticity: 207,000 MPa, Poisson ratio: 0.3; Pinion torque: 1500 Nm; All gears have optimized tooth root fillets.

Numbers of teeth	Pinion	14	15	16	17	18	19	20	21	22	23	24	25	26	27	28
	Gear	28	30	32	34	36	38	40	42	44	46	48	50	52	54	56
Module, mm		7.143	6.667	6.25	5.882	5.556	5.263	5.000	4.762	4.545	4.348	4.167	4.000	3.846	3.704	3.571
Tooth thickness ratio		0.518	0.520	0.522	0.524	0.526	0.527	0.529	0.531	0.532	0.534	0.536	0.537	0.538	0.540	0.541
Drive pressure angle,°		42.0	39.1	36.6	34.5	32.7	31.1	29.7	28.3	27.0	25.8	24.8	23.9	23.0	22.3	21.5
Asymmetry factor		1.300	1.245	1.203	1.172	1.148	1.128	1.112	1.097	1.084	1.073	1.064	1.057	1.049	1.044	1.038
Nominal drive contact ratio		1.11	1.19	1.27	1.35	1.43	1.51	1.59	1.67	1.75	1.83	1.91	1.99	2.07	2.15	2.23
Effective drive contact ratio		1.24	1.33	1.42	1.51	1.60	1.69	1.78	1.86	1.96	2.04	2.13	2.22	2.31	2.40	2.49
Specific sliding velocities		0.279	0.292	0.301	0.309	0.316	0.321	0.322	0.330	0.334	0.338	0.340	0.343	0.345	0.347	0.349
Bearing load, N		40368	38601	37417	36448	35647	35037	34890	34072	33673	33320	33045	32839	32616	32422	32247
Load sharing factor		1.0	1.0	1.0	1.0	1.0	1.0	1.0	1.0	1.0	0.743	0.692	0.674	0.644	0.630	0.620
Contact ratio type		–	Low		–		Medium		–		Transitional		–		High	–
Selected gear set number		–	1		–		2		–		3		–		4	–

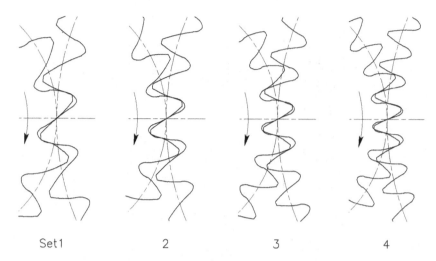

FIGURE 5.6
Selected set gear meshes at the same scale; arrows indicate the driving pinion torque direction. (Kapelevich A.L. and Y.V. Shekhtman. Analysis and optimization of contact ratio of asymmetric gears, *Gear Technology*. March/April 2017, 66–71. With permission.)

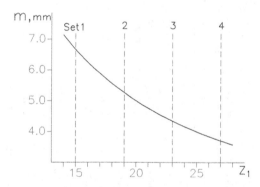

FIGURE 5.7
Gear sets' operating module.

module m is reducing to maintain the constant center distance a_w. The tooth thickness ratio TTR is slightly increasing, keeping equal specific sliding velocities H_{sd} and H_{td}, and the pinion and gear relative tooth tip thicknesses $m_{a1,2}$. The drive flank pressure angle α_{wd} is getting lower, and the nominal and effective contact ratios ε_{ad} and ε_{ade} are increasing, providing constant mesh efficiency E. As a result of the drive flank pressure angle reduction, the bearing load F is getting lower, but the equalized specific sliding velocities H_{sd} and H_{td} are increasing because of the increased contact ratios.

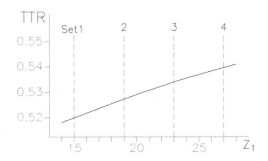

FIGURE 5.8
Tooth thickness ratio.

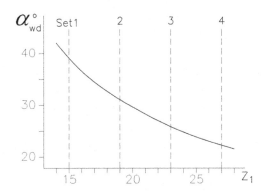

FIGURE 5.9
Drive flank pressure angle.

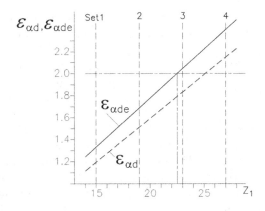

FIGURE 5.10
Nominal and effective drive flank contact ratios.

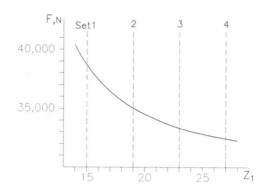

FIGURE 5.11
Bearing load.

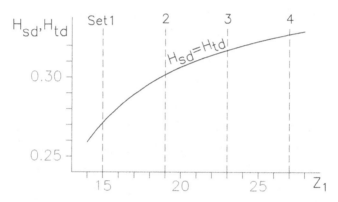

FIGURE 5.12
Drive flank specific sliding velocities.

The results of the selected gear set stress analysis and effective contact ratio, load sharing factor, and transmission error calculations under the different driving torque values are shown in Table 5.4.

Figure 5.13 presents charts of the effective contact ratio vs. the pinion torque. The effective contact ratios are gradually growing with an increase of the pinion torque, because of increasing of bending and contact tooth deflections.

The load sharing factors of the sets 1 and 2 (Figure 5.14) are equal to 1.0 because the effective contact ratios of these gear are less than 2.0 within the given pinion torque range. The gear set 3 with a transitional drive contact also has the load sharing factor equal to 1.0 until the pinion torque exceeds a value (shown as a vertical dashed line in Figures 5.13–5.16 and 5.18). Then the single/double tooth pair contact when the effective contact ratio < 2.0 becomes the double/triple tooth pair contact when the effective contact ratio ≥ 2.0 and the load sharing factor is reducing below 1.0. The gear set 4

TABLE 5.4

Selected Gear Set Analysis Results with Different Driving Torques

Gear set 1 – low drive contact ratio: $\varepsilon_{ad} = 1.194$, $z_1 = 15$, $z_2 = 30$, m = 6.667 mm, $\alpha_{wd} = 39.0°$, $\alpha_{wc} = 15.0°$

Pinion torque, Nm	250	500	750	1000	1250	1500	1750	2000
Effective drive contact ratio	1.25	1.28	1.29	1.31	1.32	1.33	1.34	1.35
Load sharing factor	1.0	1.0	1.0	1.0	1.0	1.0	1.0	1.0
Pinion bending stress, MPa	53.7	108	161	215	269	322	376	430
Gear bending stress, MPa	52.9	106	159	212	264	317	370	423
Contact stress, MPa	714	921	1115	1288	1433	1570	1695	1812
Transmission error, μm	3.1	6.1	8.8	11.3	13.7	16.1	18.4	20.8

Gear set 2 – medium drive contact ratio: $\varepsilon_{ad} = 1.512$, $z_1 = 19$, $z_2 = 38$, m = 5.263 mm, $\alpha_{wd} = 31.1°$, $\alpha_{wc} = 15.0°$

Pinion torque, Nm	250	500	750	1000	1250	1500	1750	2000
Effective drive contact ratio	1.59	1.63	1.65	1.67	1.68	1.69	1.72	1.74
Load sharing factor	1.0	1.0	1.0	1.0	1.0	1.0	1.0	1.0
Pinion bending stress, MPa	65.0	130	195	260	325	390	455	520
Gear bending stress, MPa	66.9	134	201	267	334	401	468	535
Contact stress, MPa	717	930	1132	1311	1464	1601	1726	1845
Transmission error, μm	2.9	5.5	8.2	10.6	12.9	15.1	17.2	19.3

(Continued)

TABLE 5.4 (Continued)

Gear set 3 – transitional drive contact ratio $\varepsilon_{ad} = 1.831$, $z_1 = 23$, $z_2 = 46$, $m = 4.348$ mm, $\alpha_{wd} = 25.8°$, $\alpha_{wc} = 15.0°$

Pinion torque, Nm	250	500	750	1000	1250	1500	1750	2000
Effective drive contact ratio	1.95	1.99	2.01	2.02	2.03	2.04	2.06	2.08
Load sharing factor	1.0	1.0	0.944	0.848	0.783	0.748	0.731	0.722
Pinion bending stress, MPa	76.8	153	216	258	301	353	412	470
Gear bending stress, MPa	81.4	162	228	271	314	359	408	458
Contact stress, MPa	740	962	1143	1260	1348	1436	1528	1622
Transmission error, µm	2.6	5.0	7.3	6.8	6.2	5.8	5.9	6.3

Gear set 4 – high drive contact ratio $\varepsilon_{ad} = 2.149$, $z_1 = 27$, $z_2 = 54$, $m = 3.704$ mm, $\alpha_{wd} = 22.3°$, $\alpha_{wc} = 15.0°$

Pinion torque, Nm	250	500	750	1000	1250	1500	1750	2000
Effective drive contact ratio	2.25	2.3	2.33	2.36	2.38	2.40	2.42	2.44
Load sharing factor	0.655	0.648	0.643	0.639	0.636	0.634	0.632	0.630
Pinion bending stress, MPa	70.0	140	210	280	350	420	490	560
Gear bending stress, MPa	65.2	130	196	261	326	392	457	522
Contact stress, MPa	661	820	974	1121	1253	1366	1470	1572
Transmission error, µm	1.5	2.9	4.2	5.5	6.7	7.8	8.9	9.9

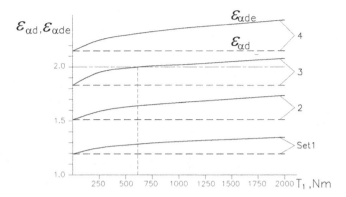

FIGURE 5.13
Effective contact ratio vs. pinion torque; dashed lines indicate the nominal contact ratio values. (Kapelevich A.L. and Y.V. Shekhtman. Analysis and optimization of contact ratio of asymmetric gears, *Gear Technology*. March/April 2017, 66–71. With permission.)

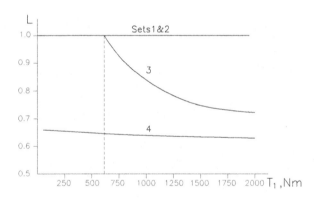

FIGURE 5.14
Load sharing factor vs. pinion torque. (Kapelevich A.L. and Y.V. Shekhtman. Analysis and optimization of contact ratio of asymmetric gears, *Gear Technology*. March/April 2017, 66–71. With permission.)

with a high drive contact ratio has the constant double/triple tooth pair contact and lowest load sharing factor, which is slightly reducing with an increase of the pinion torque.

Figures 5.15a and 5.15b show the tooth root bending stress charts of the pinion and gear. The root bending stresses of sets 1, 2, and 4 are increasing in direct proportion to the pinion torque. The gear set 3 pinion and gear tooth root bending stresses behave the same until the increasing pinion torque converts the single/double tooth pair contact into the double/triple tooth pair contact. Then the bending stress growth noticeably slows down.

The tooth flank contact stresses (Figure 5.16) of the sets 1, 2 and, 4 are growing proportionally to the square root of the pinion torque. But the gear

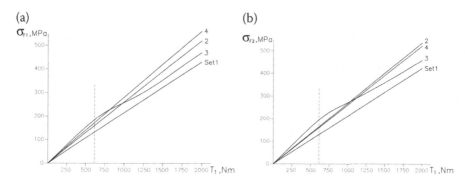

FIGURE 5.15
Root bending stress charts vs. pinion torque; (a) pinion, (b) gear. (Kapelevich A.L. and Y.V. Shekhtman. Analysis and optimization of contact ratio of asymmetric gears, *Gear Technology*. March/April 2017, 66–71. With permission.)

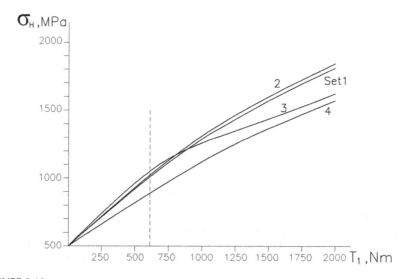

FIGURE 5.16
Contact stress vs. pinion torque. (Kapelevich A.L. and Y.V. Shekhtman. Analysis and optimization of contact ratio of asymmetric gears, *Gear Technology*. March/April 2017, 66–71. With permission.)

set 3 tooth contact stress growth is slowing down when the increasing pinion torque makes the effective drive contact ratio $\varepsilon_{ade} \geq 2.0$, and the tooth load is shared in the double/triple tooth pair contact.

Figure 5.17 presents the transmission error charts vs. the pinion rotation angle. The transmission error variation of the sets 1, 2, and 4 is increasing with the growing pinion torque, but its shape of the chart remains similar. The set 3 transmission error chart behaves differently. At the beginning

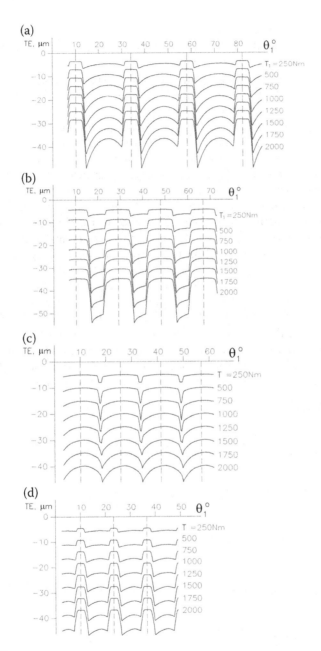

FIGURE 5.17
Transmission error charts; (a): gear set 1, (b): gear set 2, (c): gear set 3, (d): gear set 4. (Kapelevich A.L. and Y.V. Shekhtman. Analysis and optimization of contact ratio of asymmetric gears, *Gear Technology*. March/April 2017, 66–71. With permission.)

until the pinion torque is relatively low, the transmission error variation growing proportionally to the torque, but when the effective contact ratio becomes ≥ 2.0 and the double/triple tooth pair contact occurs, it is getting lower till reaching its minimum, then it is growing again. Its shape is changing during the transition from the single/double tooth pair contact to the double/triple tooth pair contact.

Figure 5.18 presents the charts of the transmission error variation vs. pinion torque for all four gear sets. The set 4 with the high drive contact ratio gears naturally has a transmission error lower than the sets 1 and 2 with the low and medium drive contact ratio gears. However, the set 3 with a transitional drive contact ratio has a transmission error variation minimum that is lower than for the gear set 4 at the same pinion torque. The transitional drive contact ratio gears can be an optimal solution for the low noise and vibration gear drives if the gear parameters are defined to have a transmission error variation minimum in the middle of the operating torque range.

A transitional nominal contact ratio value required for achieving an effective contact ratio approximately equal 2.0 is dependent on an operating load for selected gear materials. For the case hardened carburized gear steels with Modulus of elasticity 200,000–210,000 MPa and Poisson ratio ~0.3, a transitional nominal contact ratio may vary in a range of 1.80–1.95. The Modulus of elasticity and Poisson ratio of powder metal (PM) gear alloys are lower than for the gear steels. Both the Modulus of elasticity E

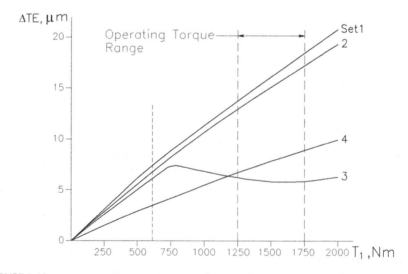

FIGURE 5.18

Transmission error variation vs. pinion torque. (Kapelevich A.L. and Y.V. Shekhtman. Analysis and optimization of contact ratio of asymmetric gears, *Gear Technology*. March/April 2017, 66–71. With permission.)

and the Poisson ratio of the PM alloy depend on its density and can be empirically calculated as following [47]:

$$E = E_0 \cdot \left(\frac{\rho}{\rho_0}\right)^{3.4} \tag{5.30}$$

and

$$\nu = \left(\frac{\rho}{\rho_0}\right)^{0.16} \cdot (1 + \nu_0) - 1, \tag{5.31}$$

where E_0, ν_0, ρ_0 are Modulus of elasticity, Poisson ratio, and density of solid material, ρ is density of the PM alloy. For solid steel $E_0 = 206$ GPa, $\nu_0 = 0.3$ and $\rho_0 = 7.86$g/cc.

Because of the lower Modulus of Elasticity and Poisson ratio, the bending and contact tooth deflections of the PM gears are larger compared to the case hardened steel gears with the same macrogeometry under the similar operating load and with the transitional contact ratio of the PM gears can be reduced to 17.7–1.85. Table 5.5 presents gear parameters of six symmetric tooth gear sets with different numbers of teeth [48]. These gear sets are optimized to satisfy the following conditions: the center distance is 150 mm; gear ratio is 2:1; pinion and gear face widths are 32 mm; tooth tip

TABLE 5.5

Comparable Symmetric Gear Set Parameters

Gear set number		1	2	3	4	5	6
Numbers of teeth	Pinion	17	18	21	23	25	19
	Gear	34	36	42	46	50	38
Module, mm		5.882	5.556	5.263	4.762	4.348	4.000
Pressure angle,°		21.5	22.3	23.0	24.1	25.00	25.5
Pitch diameter	Pinion	100.000	100.000	100.000	100.000	100.000	100.000
(PD), mm	Gear	200.000	200.000	200.000	200.000	200.000	200.000
Tooth tip diameter, mm	Pinion	115.49	114.52	112.16	110.85	109.84	113.62
	Gear	213.72	213.00	211.14	209.94	209.18	212.30
Tooth root	Pinion	84.98	86.01	88.14	89.27	90.26	86.74
diameter, mm	Gear	182.96	184.27	186.90	188.28	189.44	185.14
Tooth thickness at	Pinion	9.942	9.355	7.999	7.318	6.624	8.920
PD, mm	Gear	8.537	8.098	6.961	6.347	5.943	7.614
Nominal contact ratio		1.84	1.82	1.80	1.77	1.74	1.71
Effective contact ratio at 1500 Nm pinion torque		2.00	2.00	2.00	2.00	2.00	2.00
Gear mesh efficiency, %		98.72	98.81	98.88	99.01	99.10	99.19

thicknesses are about 0.30·module; effective contact ratio at the 1500 Nm pinion torque is equal to 2.0; assumed average friction coefficient is 0.05; pinion and gear material is Höganäs Astaloy Mo 0.25%C with the Modulus of Elasticity = 160,000 MPa and Poisson ratio = 0.28. The operating load range of all gear sets lays between 1500 and 2500 Nm of driving pinion torque. All gears have the optimized tooth root fillets.

Overlays of the pinion and gear tooth profiles of the analyzed six sets are shown in Figure 5.19.

Tables 5.6 and 5.7 present the effective contact ratio, load sharing, pinion and gear bending and contact stresses, and transmission error variation of the analyzed six gear sets under different pinion torques.

The effective contact ratio ε_{ae} versus the pinion torque chart is presented in Figure 5.20. When the pinion torque is zero, the effective contact ratio values are equal to the nominal contact ratio values because tooth deflections are zero. Increasing pinion torque T_1 increases the effective contact ratio that reaches its value of 2.0 at T_1 = 1500 Nm. Further pinion torque growth provides load sharing between two or three pairs of teeth, and the effective contact ratio $\varepsilon_{ae} > 2.0$.

Figure 5.21 shows charts of the load sharing factor L defined by Equation 5.29. At the relatively low pinion torque, the effective contact ratio $eae < 2.0$ and the load sharing factor $L = 1.0$. When the increasing pinion torque $T1$ reaches a value of about 1,500 Nm, the effective contact ratio eae becomes ≥ 2.0, the single/double tooth pair contact turns into the double/triple tooth pair contact, providing load sharing between two or three gear tooth pairs and reducing the load sharing factor $L < 1.0$. Further increase in the pinion's torque within the operating load range, additionally reduces the load sharing factor and respectively decreases the single tooth pair maximum load. This reduces tooth deflections and transmission error variation (Figure 5.22).

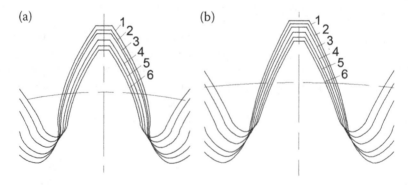

FIGURE 5.19
Gear profiles overlay; (a): pinion tooth profiles, (b): gear tooth profiles. (Kapelevich A.L. and A. Flodin. Contact ratio optimization of powder-metal gears, *Gear Solutions*. August 2017, 39–44 Contact ratio optimization of powder-metal gears. With permission.)

TABLE 5.6

Effective Contact Ratio, Load Sharing, and Transmission Error Variation

Gear set number	1	2	3	4	5	6
Pinion torque = 500 Nm						
Effective contact ratio	1.94	1.93	1.92	1.91	1.90	1.89
Load sharing, %	100	100	100	100	100	100
Transmission error variation, μm	9.2	8.9	8.6	7.9	7.2	6.5
Pinion torque = 1000 Nm						
Effective contact ratio	1.98	1.98	1.97	1.97	1.96	1.96
Load sharing, %	100	100	100	100	100	100
Transmission error variation, μm	18.5	17.8	17.1	15.8	14.4	13.1
Pinion torque = 1500 Nm						
Effective contact ratio	2.00	2.00	2.00	2.00	2.00	2.00
Load sharing, %	99.0	98.1	97.4	96.6	96.0	95.4
Transmission error variation, μm	23.0	22.5	22.0	20.9	19.8	18.6
Pinion torque = 2000 Nm						
Effective contact ratio	2.02	2.02	2.02	2.03	2.03	2.03
Load sharing, %	87.2	86.9	86.6	86.3	86.0	85.7
Transmission error variation, μm	22.1	20.6	20.0	19.0	17.9	16.9
Pinion torque = 2500 Nm						
Effective contact ratio	2.03	2.03	2.04	2.04	2.04	2.05
Load sharing, %	79.6	79.3	79.0	78.7	78.4	78.1
Transmission error variation, μm	19.7	19.2	18.7	17.6	16.5	15.5
Pinion torque = 3000 Nm						
Effective contact ratio	2.05	2.06	2.06	2.07	2.07	2.08
Load sharing, %	75.1	74.8	74.5	74.2	73.9	73.6
Transmission error variation, μm	19.1	18.6	18.1	17.0	15.9	14.9

Figures 5.23 and 5.24 present the bending and contact stress charts.

Application of the transitional contact ratio gears allows improving the gear drive performance by increasing its power density (increasing transmitted load or decreasing size and weight) and reducing transmission error variation.

5.1.2 Tooth Root Fillet Profile

Historically, gear tooth geometry improvement efforts have been concentrated on the involute flanks. Although the gear tooth root fillet is an area of maximum bending stress concentration, its profile and accuracy are marginally defined on a gear drawing. Typically, it is specified by the root diameter with a very generous tolerance and, in some cases, by the minimum fillet radius, which is not easy to measure. Required tooth root bending strength is usually provided by material and heat treatment improvement, and root surface enhancement operation, like the shot peening,

TABLE 5.7

Root Bending and Contact Stresses

Gear set number		1	2	3	4	5	6
		Pinion torque = 500 Nm					
Bending stress, MPa	Pinion	107	115	122	135	149	163
	Gear	104	110	117	130	143	156
Contact stress, MPa		1136	925	853	842	832	821
		Pinon torque = 1000 Nm					
Bending stress, MPa	Pinion	218	231	244	269	295	321
	Gear	210	222	234	258	283	307
Contact stress, MPa		1614	1304	1200	1185	1170	1155
		Pinion torque = 1500 Nm					
Bending stress, MPa	Pinion	295	315	333	372	402	434
	Gear	286	319	342	355	385	415
Contact stress, MPa		1981	1566	1426	1410	1395	1379
		Pinion torque = 2000 Nm					
Bending stress, MPa	Pinion	388	405	423	459	493	528
	Gear	373	389	406	440	472	506
Contact stress, MPa		2290	1743	1573	1550	1526	1503
		Pinion torque = 2500 Nm					
Bending stress, MPa	Pinion	486	502	521	556	591	626
	Gear	466	482	500	533	566	599
Contact stress, MPa		2562	1910	1715	1684	1652	1621
		Pinion torque = 3000 Nm					
Bending stress, MPa	Pinion	583	600	618	654	688	723
	Gear	559	575	592	626	659	692
Contact stress, MPa		2808	2070	1855	1815	1775	1735

rather than the tooth root geometry optimization. In most cases, the tooth root fillet profile is a trochoidal curve determined by the gear generating tool (hob or shaper cutter) tooth tip trajectory.

There are two general approaches for reducing the tooth root bending stress. One of them is to adjust the gear generating cutter tooth tip profile [49–51]. The most common application of such an approach is the tooling rack with the full tip radius. The second approach is to modify the tooth root fillet profile [52–58]. The most common application of this approach is to use the circular arc root fillet profile. Both of these approaches are based on a mathematical function curve fitting technique. A parabola, ellipsis [56], chain curve, so-called "bionic tooth root contour" [57,58], etc. allow reducing the root bending stress. Bending stress reduction achieved by these methods varies depending on the gear and gear cutter geometry parameters. The resulting tooth fillet profile must be checked for interference with the mating gear tooth tip. Authors of the listed above publications applied the FEA for the root stress calculation.

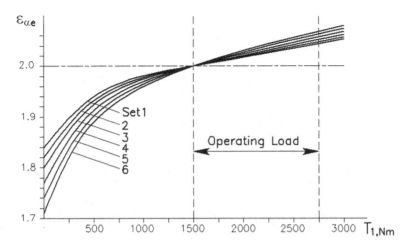

FIGURE 5.20
Effective contact ratio vs. pinion torque. (Kapelevich A.L. and A. Flodin. Contact ratio optimization of powder-metal gears, *Gear Solutions*. August 2017, 39–44 Contact ratio optimization of powder-metal gears. With permission.)

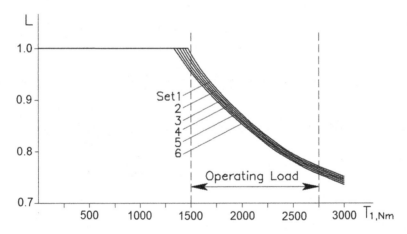

FIGURE 5.21
Load sharing factor vs. pinion torque. (Kapelevich A.L. and A. Flodin. Contact ratio optimization of powder-metal gears, *Gear Solutions*. August 2017, 39–44 Contact ratio optimization of powder-metal gears. With permission.)

5.1.2.1 Optimization Method

Direct Gear Design® optimizes the tooth root fillet profile after the complete definition of the involute flank macrogeometry parameters. The goal is to achieve a minimum of stress concentration on the tooth root fillet profile by increasing the root fillet curvature radius in the maximum stress area and even stress distribution along the large portion of the root fillet.

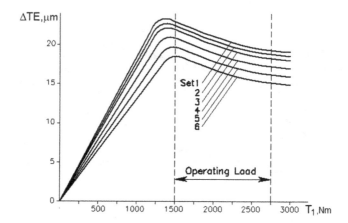

FIGURE 5.22

Transmission error variation vs. pinion torque. (Kapelevich A.L. and A. Flodin. Contact ratio optimization of powder-metal gears, *Gear Solutions*. August 2017, 39–44 Contact ratio optimization of powder-metal gears. With permission.)

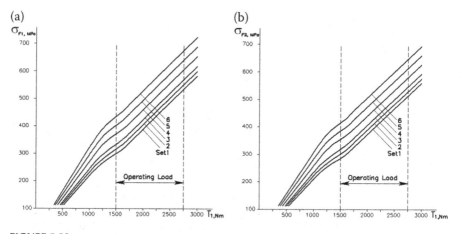

FIGURE 5.23

Bending stresses vs. pinion torque; a: pinion, b: gear. (Kapelevich A.L. and A. Flodin. Contact ratio optimization of powder-metal gears, *Gear Solutions*. August 2017, 39–44 Contact ratio optimization of powder-metal gears. With permission.)

The initial root fillet profile is a trajectory of the mating gear tooth tip in the tight (zero-backlash) engagement (see Figure 5.25). This shape of the initial root fillet completely excludes an interference with the mating gear tooth tip.

The Direct Gear Design® tooth root fillet optimization method was developed by Dr. Yuriy V. Shekhtman [59,60]. It utilizes the following calculation procedures:

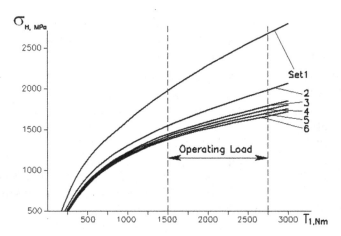

FIGURE 5.24
Contact stress vs. pinion torque. (Kapelevich A.L. and A. Flodin. Contact ratio optimization of powder-metal gears, *Gear Solutions*. August 2017, 39–44 Contact ratio optimization of powder-metal gears. With permission.)

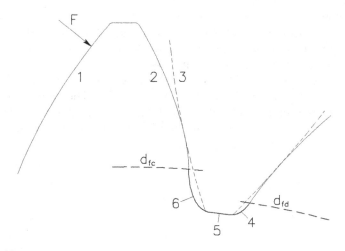

FIGURE 5.25
Generation of the initial root fillet profile as a trajectory of the mating gear tooth tip; 1: drive involute flank, 2: coast involute flank, 3: mating gear tooth in the tight (zero-backlash) engagement, 4: portion of the initial root fillet formed by the drive flank tip radius of the mating gear tooth, 5: portion of the initial root fillet formed by the tip land of the mating gear tooth, 6: portion of the initial root fillet formed by the coast flank tip radius of the mating gear tooth, d_{fd}: drive flank form circle, d_{fc}: coast flank form circle.

- Definition of a set of mathematical functions that are used to describe the optimized fillet profile. This set contains a combination of trigonometric, polynomial, and exponential functions.
- Parameters of these mathematical functions are defined during the optimization process.
- The 2D FEA with the triangle linear elements is used for stress calculation. This kind of finite element allows achieving satisfactory optimization results within reasonable calculation time.
- A random search method [61] and a modified combination of the gradient descent and ravine methods are used to define the next step in the multiparametric iteration process of the gear tooth root fillet profile optimization.

At the beginning of the root fillet optimization process, the finite element (FE) nodes are evenly distributed on the initial fillet profile. As shown in Figure 5.26, a center of the symmetric tooth root fillet A is constructed at the intersection point of the lines perpendicular to the involute flanks at the first and last points of the initial root fillet. These points are located on the form circle d_f. The FE nodes are connected to the root fillet center A by the straight lines (beams), used as the FE node movement paths during the root fillet optimization.

Figure 5.27 presents the definition of the initial location of the FE nodes for the asymmetric tooth root fillet optimization. In this case, the center points A and B are defined as the intersection points of three lines. One of them connects the center of the gear and passes the center of the initial root fillet formed by the tip land of the mating gear tooth. The other two lines are perpendicular to the involute flanks at the first and last points of the initial root fillet. Then the line AB is evenly divided by a number of points equal to

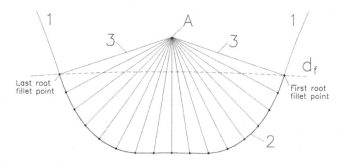

FIGURE 5.26
Symmetric tooth root fillet FE node movement paths definition; 1: involute tooth flanks, 2: initial root fillet profile, 3: lines perpendicular to the involute flanks at the first and last points of the initial root fillet that are located on the form circle d_f, A: root fillet center.

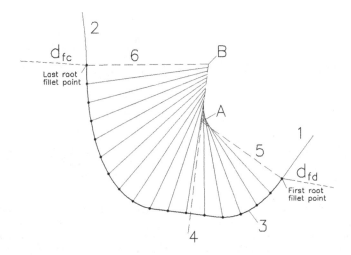

FIGURE 5.27
Asymmetric tooth root fillet FE node movement paths definition; 1: drive involute flank, 2: coast involute flank, 3: initial root fillet profile, 4: line connecting a center of the gear with a center of the initial root fillet portion that is formed by the tip land of the mating gear tooth, 5: line perpendicular to the drive involute flank at the first point of the initial root fillet, 6: line perpendicular to the coast involute flank at the last point of the initial root fillet, A and B: root fillet centers.

the number of the root fillet FE nodes and each of these points is connected to its FE node. These straight lines (beams) are used as the root fillet FE node movement paths.

The first and last finite element nodes of the initial fillet profile located on the form diameter circle are fixed during the root optimization process. The rest of the initial fillet FE nodes are moved along their beams. The root stresses are calculated for every fillet profile configuration iteration. Variable parameters of mathematical functions that describe the fillet profile for the next iteration are defined depending on the stress calculation results of the previous iterations. Both successful (resulting with a stress reduction) and unsuccessful (resulting with a stress increase) iterations define the next iteration direction.

After the specified number of iterations, indicating that further stress reduction is not achievable, the optimization process stops, resulting in the optimized fillet profile. Figures 5.28 and 5.29 illustrate the root fillet optimization and show photos of the optimized root fillet.

As a result of the symmetric tooth root fillet optimization, the tensile and compressive stresses are evenly distributed along large portions of the root fillet profile. The stress charts are shown in Figure 5.30 for the symmetric gear tooth and Figure 5.31 for the asymmetric gear tooth.

(a)   (b)

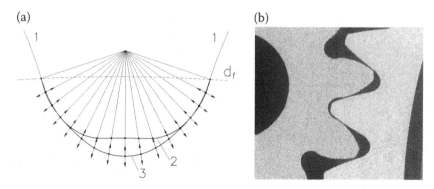

FIGURE 5.28
Optimized root fillet of the symmetric tooth gear; (a) a final position of the root fillet FE nodes; (b) a photo of the gears with the optimized root fillets; 1: involute flanks, 2: initial root fillet profile, 3: optimized root fillet profile. [(b) From Kapelevich, A.L., and Y.V. Shekhtman, Tooth fillet profile optimization for gears with symmetric and asymmetric teeth. AGMA Fall Technical Meeting, San Antonio, Texas, October 12–14, 2008, 08FTM06. With permission.]

(a)   (b)

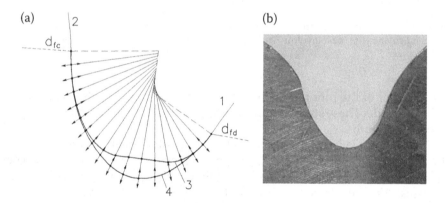

FIGURE 5.29
Optimized root fillet of the asymmetric tooth gear; (a) a final position of the root fillet FE nodes; (b) a photo of the optimized root fillet; 1: drive involute flank, 2: coast involute flank, 3: initial root fillet profile, 4: optimized root fillet profile.

Figure 5.32 demonstrates the FEA mesh of the asymmetric gear tooth with the optimized root fillet. The FE nodes at the bottom of the mesh model, shown by the black dots, are constrained in all directions and present the boundary conditions of the solid body gear. The thin rim gear tooth root optimization is considered in the next section. During the optimization process, a tooth force F is applied perpendicularly to the drive tooth flank at the highest point of single tooth contact (HPSTC) for the low and medium contact ratio spur gears or at the highest point of double tooth contact for the HCR spur gears. Typically, a tooth force F application point does not

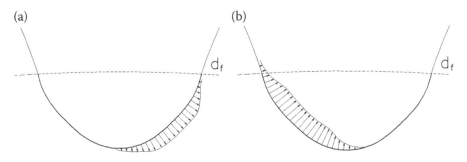

FIGURE 5.30
Tensile (a) and compressive (b) root bending stress charts along the optimized root fillet profile of the symmetric gear tooth.

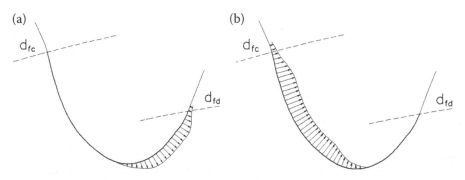

FIGURE 5.31
Tensile (a) and compressive (b) root bending stress charts along the optimized root fillet profile of the asymmetric gear tooth.

coincide with a FE node and is located between two nodes. Then it is replaced by two forces applied to those nodes. A sum of these two forces is equal to the initial single force F, and their separate values are defined in an inverse proportion to the distances from the initial single force F application point to the loaded nodes.

Figure 5.33 shows the typical tooth stress isograms and stress distribution along with the tooth profile after the root fillet optimization. These charts illustrate how bending tensile and compressive stresses are evenly distributed along large portions of the tooth root fillet.

The more the number of finite element nodes placed on the root fillet profile, the more accurate the stress calculation results, but this requires more iterations, and the root fillet profile optimization takes more time. During the optimization process, the fillet nodes cannot be moved inside the initial fillet profile because this may cause interference with the mating gear tooth tip. It makes unnecessary a checkup of the optimized root fillet for interference with the mating gear tooth tip. The second derivative at

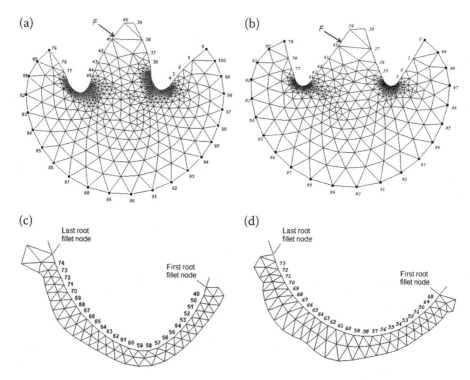

FIGURE 5.32
FEA mesh model of the asymmetric gear tooth (a) external gear and (b) internal gear and magnified optimized root fillet (c) external gear and (d) internal gear; *F*: applied tooth load; black dots in views (a) and b: boundary FE nodes constrained in all directions.

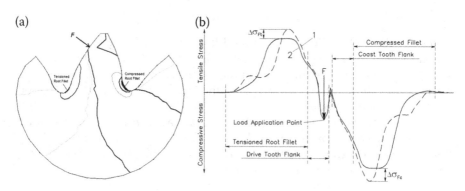

FIGURE 5.33
Asymmetric gear tooth stress isograms (a) and a stress distribution along the tooth profile (b); *F*: applied tooth load, 1: before root fillet optimization, 2: after root fillet optimization; $\Delta\sigma_{Ft}$: tensile stress reduction, $\Delta\sigma_{Fc}$: compressive stress reduction.

any point of the optimized root fillet curve must be positive to exclude its waviness.

The tensile σFt and compressive σFc stress reduction after the root fillet optimization (Figure 5.33b) depends on the shape of the initial root fillet that is a trajectory of the mating gear tooth tip in the tight (zero-backlash) engagement. If the mating gear number of teeth is low and its tip tooth thickness is small (close to the pointed tooth tip), the stress reduction after optimization is relatively low 10%–15% in comparison to the initial root fillet. If the mating gear has many teeth or it has internal teeth, and the tooth tip is large, the stress reduction after optimization could be as high as 30%–40%.

There are other possible root fillet optimization constraints. One of them is a minimum radial clearance. The optimization process tends to minimize a radial clearance to reduce the tooth root stress, making it lower than that in conventional standard gears. The concern here is that the radial clearance can be so small that the lubricant would be trapped in the root fillet space, resulting in additional hydraulic power losses and gear efficiency reduction. In this case, the required root diameter, providing the acceptable radial clearance, constrain the root fillet optimization process, and the optimized root fillet profile must be tangent to the root circle (see Figure 5.34).

Another root fillet optimization constraint is related to the manufacturability of the gears with the optimized fillet. This is typical for external gears with a low (<20) number of teeth. The root fillet optimization process tries to reduce curvature (increase radius) at the maximum stress root fillet area to minimize its concentration. For gears with a low number of teeth, it may

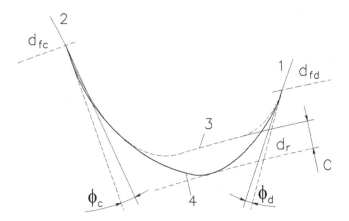

FIGURE 5.34
Tooth root fillet optimization constraints (dashed lines); 1: drive involute flank, 2: coast involute flank, 3: initial root fillet profile, 4: optimized root fillet profile, d_{fd} and d_{fc}: drive and coast flank form circle diameters, d_r: required root diameter, C: radial clearance, φ_d and φ_c: drive and coast flank undercut angles.

result in a small fillet radius near the form diameter creating the undercut. Unlike the undercut that occurs in conventional gears, this one is made for the purpose, and it does not affect the active involute flank profile. However, producing such an optimized fillet profile could be difficult or even impossible by some gear fabrication methods, such as, for example, form cutting, hobbing, etc. In order to make the root fillet profile manufacturable using these fabrication methods, the root fillet profile undercut should be limited or completely eliminated. This constrains the undercut angles φd and φc (Figure 5.34) between tangents to the initial and optimized fillet profiles at their first and last points. These additional fillet optimization constraints may partially compromise root stress reduction in comparison to the optimized fillet constrained only by the initial fillet profile.

The root fillet optimization process intends to minimize the maximum root bending stress trying to shape the root fillet for even redistribution of the tensile and compressive stresses. This stress redistribution may increase a total stress magnitude (a difference between the maximum tensile and compressive values) at some point of the root fillet. Typically, this occurs at the very bottom of the root fillet near the root circle. If this happens, a total stress magnitude becomes an additional constraint. In such cases, the root fillet optimization should simultaneously combine the bending stress and total stress magnitude minimization. Figure 5.35 presents the comparison chart of the total stress magnitude along with the root fillet profile before and after optimization. Such combined root fillet optimization allows reducing the total stress magnitude reduction of about 20% in comparison to

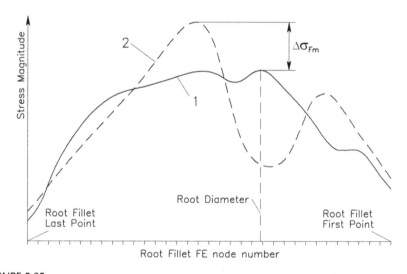

FIGURE 5.35
Total stress magnitude along the root fillet; 1: before root fillet optimization, 2: after root fillet optimization; $\Delta_{\sigma Fm}$: total stress magnitude reduction.

the initial root fillet. The necessity to simultaneously minimize the root stress and total stress magnitude minimization compromises root stress reduction in comparison to the optimized fillet constrained only by the initial fillet profile.

Table 5.8 presents the comparison of the asymmetric gear pairs designed traditionally by the rack generating method and by Direct Gear Design®. All parameters and dimensions of these gear pairs are identical, except the root fillet profiles. The traditionally designed asymmetric gear pair is generated by the asymmetric rack with a full tip radius that produces the trochoidal tooth root fillet with minimal root stresses, achievable by this gear design method. The asymmetric gear pair defined by Direct Gear Design® has an optimized root fillet, which demonstrates significant (~18%) bending stress reduction compared to the best trochoidal root fillet of the traditionally designed asymmetric gear pair.

An overlay of the pinion root fillet profiles of the traditionally and directly designed gear pair is shown in Figure 5.36.

For helical gears, the tooth root fillet should be optimized in the normal to the tooth line section.

TABLE 5.8
Tooth Root Stress Comparison

Tooth geometry definition		Rack generation		Direct Gear Design®	
Gear		Pinion	Gear	Pinion	Gear
Number of teeth		19	29	19	29
Generating Rack Coefficients	Addendum	1.000		–	
	Dedendum	1.174		–	
	Tip Radius	0.264 (Full Radius)		–	
X-shift coefficient		0.0	0.0	–	
Pressure angles (drive/coast)		35°/20°*		35°/20°*	
Module, mm		4.00	4.00	4.00	4.00
Pitch diameter (PD), mm		76.000	116.000	76.000	116.000
Tooth tip diameter, mm		84.000	124.000	84.000	124.000
Root diameter, mm		66.609	106.608	66.811	106.964
Tooth thickness at PD, mm		6.283	6.283	6.283	6.283
Face width, mm		30.0	28.0	30.0	28.0
Torque, Nm		700	1068	700	1068
Max. tensile stress, MPa		441	440	363	359
Max. compressive stress, MPa		–676	–691	–556	–565
Tensile stress reduction		–	–	17.7%	18.4%
Compressive stress reduction		–	–	17.8%	18.2%

Note:
* drive/coast flank

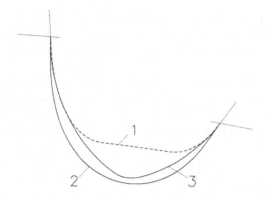

FIGURE 5.36
Pinion root fillet profile comparison; 1: trajectory of the mating gear tooth tip in the zero backlash engagement, 2: traditionally designed trochoidal root fillet, 3: root fillet optimized by Direct Gear Design®.

5.1.2.2 Thin Rim Gear Tooth Root Fillet Optimization

Thin rim gears naturally have higher bending root stress compared to the solid body gears. This makes it essential to reduce this stress by the root fillet optimization. In this case, the optimized root fillet profile and maximum root stress depend on the rim thickness and type of the gear rim design. There are several different rim design options that can be considered.

The first one is "free" rim support with its possible radial deflection. Its FEA mesh model is shown in Figure 5.37a. This is typical for gears with the spokes connecting the rim with the gear hob.

The second option of rim design is typical for gears that have the sliding fit on the internal rim surface with a very small clearance. This restrains the rim radial deflections. Its FEA mesh model is shown in Figure 5.37b.

The third type of thin rim design is when a gear is connected with a shaft by the interference fit that could be a press or shrink fit. It imposes additional hoop stress depending on the press or shrink fit interference. Its FEA mesh model is shown in Figure 5.37c. If this interference is significant, resulting in high hoop stress, or if a rim thickness is low, the root fillet optimization may not be possible.

As it is shown in Section 5.1.1.2, the application of asymmetric tooth gears in an epicyclic stage allows the contact stresses to be balanced by choosing a greater pressure angle in the sun/planet gear mesh than in the planet/ring gear mesh in order to maximize power density, reducing a gear drive size. This means that a reduction of the center distance and all gear dimensions, including the planet gear root diameter. In many cases, a planet gear is supported by a roller bearing incorporated inside the gear, using its inner diameter as a roller bearing outer race surface (Figure 5.38). Reduction of the center distance, while the transmitted torque remains constant, increases

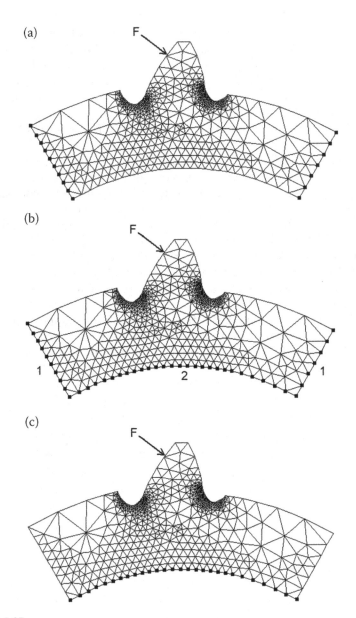

FIGURE 5.37
Thin rim gear FEA mesh models: (a): "free" rim support, the radially located FE nodes are constrained in all directions; (b): slip fit rim support, 1: radially located FE nodes are constrained in the circular direction and 2: FE nodes located on the inner radius are constrained in the radial direction; (c): interference fit rim support, the FE nodes located on the inner radius are constrained in the circular direction and are moved in the radial direction out of a center on the distance equal a half of the press fit interference.

the planet gear roller bearing load. It leads to the necessity of increasing the diameter of the roller bearing to maintain its required load capacity and life. A simultaneous reduction of the planet gear root diameter and increase of its inner diameter significantly reduce the planet gear rim thickness, increasing the tooth root bending stress. In such cases, tooth root fillet optimization is essential to keep root stresses within an acceptable level [62].

Figure 5.39 shows the planet gear tooth loading. Teeth of the planet gear are loaded at both flanks by forces applied from the sun and ring gears. In an asymmetric epicyclic stage, the tangent force components are identical (Figure 5.38). However, the normal force F_s of the sun/planet gear contact is greater than the normal force F_r of the planet/ring gear contact by a ratio equal to the asymmetry factor K. Besides, because of the tooth asymmetry and different contact ratios at the opposite tooth flanks, the HPSTC of the sun/planet gear contact is located significantly closer to the tooth tip compared to the planet/ring gear contact. For these reasons, the sun/planet gear contact force results in higher root bending stress, and this loading condition (Figure 5.39a) is preferable for the root fillet optimization.

A relative angular position of the planet gear teeth and supporting bearing roller is changing with the gear rotation. Two of the relative angular positions are considered for the planet gear tooth root fillet optimization. The first

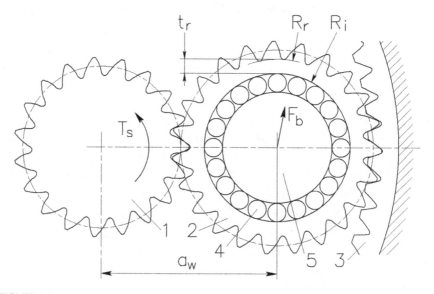

FIGURE 5.38

Epicyclic stage with the thin rim planet gear; 1: sun gear, 2: planet gear, 3: ring gear, 4: bearing rollers, 5: planet gear shaft, T_s: sun gear torque, a_w: center distance, F_b: planet gear bearing load, R_r: planet gear root radius, R_i: planet gear inner radius, t_r: rim thickness. (Kapelevich A.L. and Y.V. Shekhtman. Root fillet optimization of thin rim planet gears with asymmetric teeth. Gear Solutions, March 2018, 52–55. With permission.)

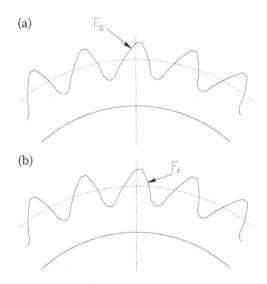

FIGURE 5.39
Planet gear tooth loading; (a): by a force *Fs* applied from the sun gear, (b): by a force *Fr* applied from the ring gear. (Kapelevich A.L. and Y.V. Shekhtman. Root fillet optimization of thin rim planet gears with asymmetric teeth. *Gear Solutions*, March 2018, 52–55. With permission.)

one (Figure 5.40a) is when the bearing roller is in line with the middle of the loaded planet gear tooth thickness. The second one (Figure 5.40b) is when the loaded planet gear tooth is between two supporting bearing rollers.

Figure 5.41 represents the thin rim planet gear FEA mesh models of two relative angular positions of the loaded gear tooth and supporting bearing rollers. The FE nodes at the inner gear diameter are constrained in the circular direction, simulating the bearing roller contact points.

Figures 5.42 and 5.43 show the planet gear tooth stress isograms and stress distribution along with the tooth profile. Since the load from the sun gear is selected as the load application condition for the root fillet optimization, it results in even stress distribution along large portions of the root fillet. But when a load from the ring gear is applied to the opposite tooth flank, a stress distribution along the root fillet is not even. Nevertheless, the maximum tensile stress, in this case, is just a bit greater compared to when a load is applied from the sun gear, but the maximum compressive stress is lower.

Table 5.9 presents the geometric data and torque for the epicyclic gear stage with asymmetric tooth gears.

Table 5.10 shows the maximum values of the planet gear root stress for the different bearing roller positions, load applications, and relative rim tooth thickness (rim thickness/module) values. Figure 5.44. provides the charts of the maximum root stresses vs. the relative rim thickness. They indicate that when a loaded tooth is in line with the bearing roller, the root stresses are generally greater than when the loaded tooth is between two rollers. If the

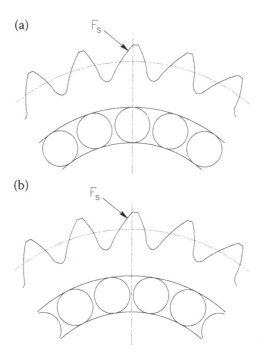

FIGURE 5.40
Planet gear tooth load and bearing roller positions: (a): tooth is in line with a roller, (b): tooth is between two rollers; F_s: load applied from the sun gear. (Kapelevich A.L. and Y.V. Shekhtman. Root fillet optimization of thin rim planet gears with asymmetric teeth. *Gear Solutions*, March 2018, 52–55. With permission.)

rim thickness exceeds the 3.5× module, the tooth root stresses do not depend on the rim thickness and practically are equal to the solid body gear stresses. If the rim thickness is less than 3.5× module, tooth root stresses grow exponentially as rim thickness is reducing. When the rim thickness that is less than 2.0× module, the maximum root stresses are getting very high. Such thin rim planet gear design is typically not recommended.

Table 5.11 shows how the planet gear root and inner diameters depend on rim thickness when the tooth root fillet is optimized. Figure 5.45 demonstrates the optimized tooth root fillet profiles and internal diameters of the planet gear for different rim thicknesses.

5.1.2.3 Optimization of Tooth Root Generated with Protuberance Hob

In the previous sections, tooth root fillet optimization is considered assuming that the involute flanks and root fillets are processed (machined) simultaneously. However, many applications require high gear tooth flank accuracy and high load capacity as well as low gear production costs. To satisfy these requirements, the tooth involute flanks, tips, and root fillets

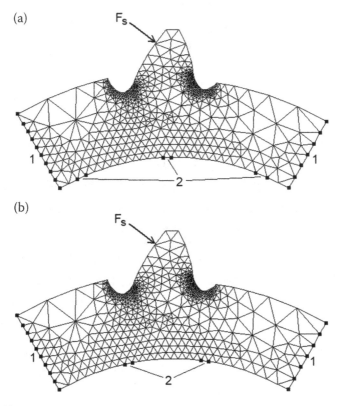

FIGURE 5.41
Thin rim planet gear FEA mesh models with the bearing roller positions: (a)tooth is in line with bearing roller, (b): tooth is between two bearing rollers; 1: FE nodes constrained in the circular direction, 2: radially constrained FE nodes; F_s: load applied from the sun gear. (Kapelevich A.L. and Y.V. Shekhtman. Root fillet optimization of thin rim planet gears with asymmetric teeth. *Gear Solutions*, March 2018, 52–55. With permission.)

are processed separately. The gear blank is machined by the topping protuberance hob, which finalizes the root fillet, tooth tip diameter, and chamfers but leaves a stock for tooth flank grinding. Then, after gear heat treatment (carburizing + case hardening) and, in some cases, shot peening of the tooth root, the tooth flanks are processed by the highly productive generating grinding, removing the grinding stock. As a result, the compressive residual stress in the root fillet, developed during case hardening, is retained. This fabrication sequence is most typical for automotive transmission gears.

Since tooth root load capacity is the main contributor to gear transmission performance, the reduction of tooth bending stress concentration in the gears generated with a protuberance hob is critical [63]. This section demonstrates the construction and optimization of the tooth root fillet while leaving the required grinding stock on the involute tooth flanks.

(a)

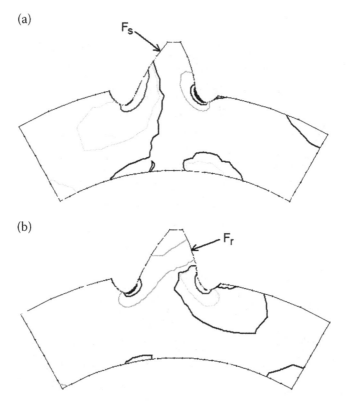

(b)

FIGURE 5.42
Thin rim planet gear FEA mesh models with the bearing roller positions: (a): tooth is in line with a bearing roller, (b): tooth is between two bearing rollers; 1: FE nodes constrained in the circular direction, 2: radially constrained FE nodes; F_s: load applied from the sun gear, F_r: load applied from the ring gear. (Kapelevich A.L. and Y.V. Shekhtman. Root fillet optimization of thin rim planet gears with asymmetric teeth. *Gear Solutions*, March 2018, 52–55. With permission.)

Figure 5.46 shows the construction of the initial root fillet of the asymmetric tooth. The Direct Gear Design® tooth geometry calculation defines the final (after grinding) drive 1 and coast 2 involute tooth flanks and also the trajectory of the mating gear tooth tip in zero backlash mesh 3. Then the preliminary (before grinding) drive 4 and coast 5 involute flanks are constructed equidistant to the final tooth flanks 1 and 2 respectively, with the offsets Δ_d and Δ_c that are the drive and coast flank grinding stocks. The interim drive 6 and coast 7 root fillet flanks are for transition between the preliminary flanks 4 and 5 and the root fillet. They intersect the preliminary flanks, providing drive and coast flank grinding stocks Δ_d and Δ_c at the lowest points of the active involute flank contact with the mating gear tooth. These points lay on the drive and coast flank lowest contact point diameters d_{lpd} and d_{lpc}. These interim fillet profiles are typically involute, formed by the straight profiles of

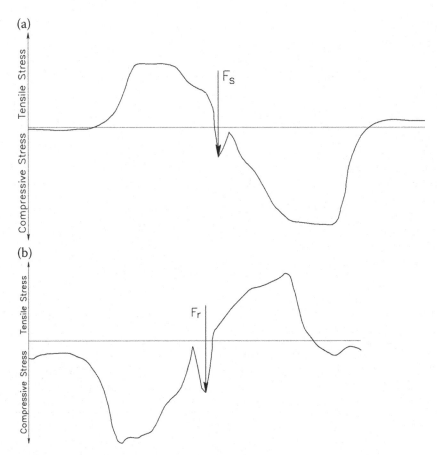

FIGURE 5.43
Stress distribution along the planet gear tooth profile; a: load F_s applied from the sun gear, b: load F_r applied from the ring gear. (Kapelevich A.L. and Y.V. Shekhtman. Root fillet optimization of thin rim planet gears with asymmetric teeth. *Gear Solutions*, March 2018, 52–55. With permission.)

the protuberance hob. The interim fillet profiles are designed to be machined by hobbing and later removed by tooth flank grinding after heat treatment.

Depending on parameters of the drive and coast flanks near the tooth root, including the lowest contact point diameters d_{lpd} and d_{lpc}, flank grinding stocks Δ_d and Δ_c, and interim fillet profile angles γ_d and γ_c, the initial tooth root fillet can have different shapes. Figure 5.46 shows two initial root fillet options. One of them is the single circular arc 8a, and another, 8b, presents 3 tangent arcs. In both cases, the initial root fillet must lay below the trajectory of the mating gear tooth tip 3 to exclude root-tip interference. A center of the middle arc of the 3-arc fillet 9 coincides with the gear center. The single circular arc 8a and the arcs at the ends of the 3-arc fillet 8b should be tangent to the respective interim root fillet flanks 6 and 7,

TABLE 5.9

Epicyclic Gear Stage Geometric Data

Gear	Sun	Planet		Ring
Number of teeth	19	23		65
Module (m)	5.00	5.00		5.00
Drive pressure angle	36°	36°	19°	19°
Coast pressure angle	19°	–	–	36°
Pitch diameter (PD)	95.00	115.00		325.00
Tooth tip diameter	105.28	125.17		317.84
Root diameter	83.69	103.52		336.04
Tooth thickness at PD	7.94	7.76		7.94
Face width	40.0	37.0		34.0
Center distance		105.00		
Drive contact ratio		1.25	1.60	
Torque per one mesh, Nm	1000	1210		3421

TABLE 5.10

Epicyclic Gear Stage Geometric Data

Root stress, MPa	Relative rim thickness	Tooth is in line with a bearing roller		Tooth between two bearing rollers	
		Loaded by sun gear	Loaded by ring gear	Loaded by sun gear	Loaded by ring gear
Tensile	4.0	273	295	261	281
	3.5	276	300	262	282
	3.0	282	308	263	284
	2.5	296	323	265	292
	2.0	330	359	270	312
	1.5	415	446	291	353
	1.0	623	662	367	468
Compressive	4.0	−398	−300	−414	−286
	3.5	−402	−306	−416	−290
	3.0	−407	−313	−423	−296
	2.5	−417	−331	−439	−308
	2.0	−456	−371	−468	−328
	1.5	−552	−456	−529	−360
	1.0	−763	−688	−730	−437

and simultaneously tangent to final involute profiles 1 and 2 near the tooth root. Although for practical purpose, there should be a small amount of undercut at an intersection point of the final involute tooth flank and the initial (and then optimized) root fillet (Figure 5.47) to accommodate

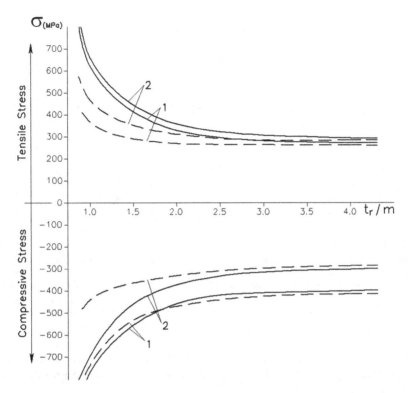

FIGURE 5.44
Maximum planet gear tooth root stress charts; 1: load applied from the sun gear, 2: load applied from the ring gear; solid lines: the loaded tooth is in line with a roller; dashed lines: the loaded tooth is between two rollers. (Kapelevich A.L. and Y.V. Shekhtman. Root fillet optimization of thin rim planet gears with asymmetric teeth. *Gear Solutions*, March 2018, 52–55. With permission.)

TABLE 5.11

Root and Internal Diameters of Thin Rim Planet Gears

20.00 (4.0 × m)	103.54	63.54
17.50 (3.5 × m)	103.59	68.59
15.00 (3.0 × m)	103.64	73.64
12.50 (2.5 × m)	103.70	78.70
10.00 (2.0 × m)	103.80	83.80
7.50 (4.0 × m)	103.99	88.99
5.00 (1.0 × m)	104.26	94.26

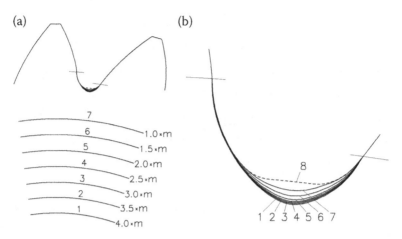

FIGURE 5.45

a: optimized tooth root fillet profiles and related internal diameters of the planet gear; 1: $t_r = 4.0 \times m$, 2: $t_r = 3.5 \times m$, 3: $t_r = 3.0 \times m$, 4: $t_r = 2.5 \times m$, 5: $t_r = 2.0 \times m$, 6: $t_r = 1.5 \times m$, 7: $t_r = 1.0 \times m$, 8: trajectory of the sun gear tooth tip in the zero-backlash engagement; b: magnified root fillet profiles. (Kapelevich A.L. and Y.V. Shekhtman. Root fillet optimization of thin rim planet gears with asymmetric teeth. *Gear Solutions*, March 2018, 52–55. With permission.)

manufacturing tolerances and heat treatment distortion to guarantee an exit for the grinding wheel without creating a step between the ground tooth flank and hobbed root fillet.

The definition of the drive and coast flank interim fillet profile angles γ_d and γ_c, considering the manufacturability of the optimized root fillet, described in the tooling design section of Chapter 9.

The only difference in the optimization of the tooth root, generated with the protuberance hob, is that the initial root fillet is defined to provide the required flank grinding stocks. Otherwise, the root optimization technique remains the same as described in Section 5.1.2.1.

Table 5.12 presents a comparison of two asymmetric gear pairs with identical flank geometry and different root fillets – a conventional trochoidal fillet generated by a protuberance hob with a full radius tooth tip, and an optimized fillet generated by a protuberance hob with a special shape tooth tip

5.1.2.4 Benefits of Tooth Root Fillet Optimization

Figure 5.48 presents a comparison of different tooth root fillet profiles of the standard proportions spur symmetric gear tooth. The involute flanks, face widths, tooth load, and application point are identical, but the root fillet profiles are different. Table 5.13 shows the results of the FEA stress calculation and other root fillet parameters. Calculation results for the root fillet profile #1 generated by the standard 20° pressure angle rack profile are

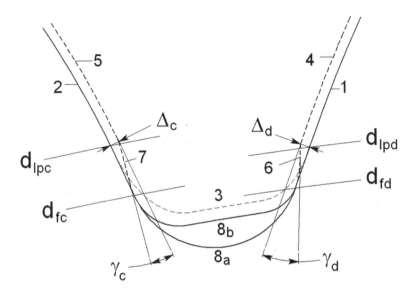

FIGURE 5.46
Initial construction of root fillet; 1 and 2: drive and coast final involute flanks after grinding; 3: trajectory of mating gear tooth tip in zero backlash mesh; 4 and 5: preliminary drive and coast involute flanks before grinding, 6 and 7: drive and coast interim involute fillet flanks that will be removed by tooth flank grinding; 8a and 8b: initial tooth root fillet options, formed by a single arc (a) and by three arcs (b); Δ_d and Δ_c: drive and coast flank grinding stocks; γ_d and γ_c: drive and coast flank interim fillet profile angles; d_{lpd} and d_{lpc}: drive and coast flank lowest contact point diameters; d_{fd} and d_{fc}: drive and coast flank form diameters. (Kapelevich A.L. and Y.V. Shekhtman. Optimization of asymmetric gear tooth root generated with protuberance hob. *Gear Solutions*, June 2020, 32–37. With permission.)

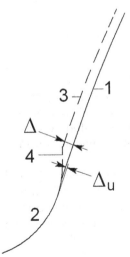

FIGURE 5.47
Small undercut Δ_u in the transition point between the final involute flank 1 and initial root fillet 2; 3: preliminary involute flank before grinding, 4: interim fillet flank; Δ: flank grinding stock. (Kapelevich A.L. and Y.V. Shekhtman. Optimization of asymmetric gear tooth root generated with protuberance hob. *Gear Solutions*, June 2020, 32–37. With permission.)

TABLE 5.12

Comparison of Gears with Conventional Trochoidal and Optimized Root Fillets Generated by Protuberance Hobs

Gear	Pinion		Gear	
Number of teeth	23		41	
Normal module, mm	3.000			
Normal pressure angle	30.00°/22.00°*			
Helix angle	30.00°			
Pitch diameter (PD), mm	79.674		142.028	
Tooth tip diameter, mm	86.475		148.424	
Root fillet profile	Conventional Trochoidal**	Optimized	Conventional Trochoidal**	Optimized
Root diameter, mm	71.654	71.448	133.714	133.520
Tooth thickness at PD, mm	4.744		4.515	
Face width, mm	40.00		38.00	
Drive flank torque, Nm	700		1248	
Root tensile stress, MPa	391	351	396	356
Tensile stress reduction	–	10.2%	–	10.1%

Notes:
 * Drive/coast flank
 ** Generated by a full tip radius protuberance hob

considered the 100% benchmark values. Parameters of other root fillet profiles are defined relative to the profile #1 parameters.

Table 5.13 presents the root fillet profile comparison that indicates considerable root stress concentration reduction provided by the root fillet optimization. The gear tooth with the optimized root fillet has larger curvature radius Rf, and smaller distance H and root clearance C. It also has the lowest maximum tensile stress, which is evenly distributed along the large portion of the root fillet profile. Other root fillet profiles have significantly higher, sharply concentrated maximum stress.

Analysis of the root fillet optimization results indicates that the optimized fillet profile practically does not depend on the force value and its application point along the involute flank, except in the case when the application point is located close to the root fillet. Such load application should not be considered for root fillet optimization, because it induces relatively low root fillet tensile stress.

(a)   (b)

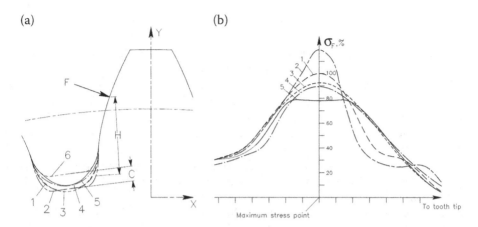

FIGURE 5.48

Root fillet profile comparison: (a): gear tooth with different fillet profiles, (b): stress chart along the fillet; 1: root fillet profile generated by the standard coarse pitch rack with the tip radius $0.3 \times m$ (or $0.3/DP$); 2: root fillet profile generated by the standard fine pitch rack with the tip radius equal to zero; 3: root fillet profile generated by the full tip radius rack; 4: circular root fillet profile; 5: optimized root fillet profile; 6: trajectory of the mating gear tooth tip in tight (zero backlash) mesh; F: applied load; H: radial distance between load application and maximum stress points; C: radial clearance; R_f: root fillet curvature radius at the maximum stress point; σ_F: tensile stress. (From Kapelevich, A.L., and Y.V. Shekhtman, Tooth fillet profile optimization for gears with symmetric and asymmetric teeth. *AGMA Fall Technical Meeting*, San Antonio, Texas, October 12–14, 2008, 08FTM06. With permission.)

TABLE 5.13

Root Fillet Profile Comparison (Figure 5.48)

	Rack with tip radius $R = 0.3$ m	Rack with tip radius $R = 0$	Rack with full tip radius	Circular root fillet profile	Optimized root fillet
Root profile #	1	2	3	4	5
Curvature radius, R_f, %	100	58	118	121	273
Distance, H, %	100	103	100	88	82
Radial clearance, C, %	100	100	118	79	76
Max. tensile stress, σ_{Fmax}, %	100	119	90	88	78

The main benefit of the root fillet optimization is the bending stress concentration reduction. If the gear drive load capacity depends on the root bending stress, the root fillet profile optimization increases the load capacity proportionally to the root stress reduction. However, quite often, the gear drive load capacity is limited by the tooth surface durability, defined by pitting and scuffing resistance, which depends on the contact stress, profile sliding, and contact (flash) temperature. In this case, potential bending stress reduction provided by the fillet optimization is convertible into other gear drive characteristics' improvement.

Figure 5.49 presents charts of the bending and contact (the dashed curve) stresses calculated for the gear pairs with the standard involute flank profiles and a gear ratio $u = 1.0$. All gear pairs have the same center distance aw = 60 mm, and the face width b = 10 mm of each gear. The applied driving torque is T = 50 Nm. The numbers of teeth $z12$ vary from 12 to 75, making the module values from 5.0 to 0.8 mm to keep the constant center distance. The bending stresses are presented in two charts: the top one is for the gears with the standard (generated by the 20° pressure angle rack) trochoidal fillet profiles, and the bottom one is for the same gears, but with the optimized fillet profiles. For example, the bending stress level of 180 MPa is considered acceptable. This level is achievable for 20-tooth gears with the standard fillet or for 28-tooth gears with the optimized fillet. However, the lower module 28-tooth gear pair has a higher contact ratio and, as a result, lower contact stress. The root fillet optimization allows for trading of the potential 24% bending stress reduction for the 6% contact stress reduction by using the gears with a larger number of teeth and lower module. This 6% contact stress reduction could be used to increase gear drive life or for size and weight reduction.

Similarly, the fillet optimization makes it possible to increase the number of gear teeth and reduce their module, allowing for reduction of specific sliding velocities (Figure 5.50) because of the tooth addendum reduction and tooth tip profile angles. It also leads to increased mesh efficiency

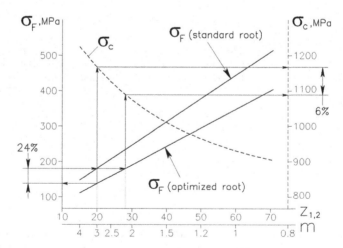

FIGURE 5.49
Conversion of potential bending stress reduction into contact stress reduction. (From Kapelevich, A.L., and Y.V. Shekhtman, Tooth fillet profile optimization for gears with symmetric and asymmetric teeth. *AGMA Fall Technical Meeting*, San Antonio, Texas, October 12–14, 2008, 08FTM06. With permission.)

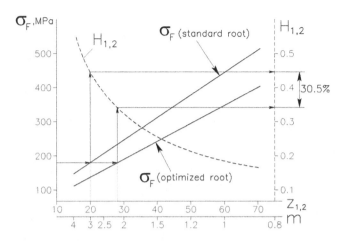

FIGURE 5.50
Conversion of potential bending stress reduction into specific sliding reduction.

(Figure 5.51), and reduced contact (flash) temperature (Figure 5.52) and scuffing probability[*] (Figure 5.53).

5.1.3 Root Stress Balance

The mating gears typically have different numbers of teeth and face widths, and are also made out of dissimilar materials. Their maximum tooth root stresses should be balanced (or safety factors should be the same) to make them equally strong [59]. This balance condition can be presented as follows:

$$|\sigma_{F\,max\,1} - C_b\sigma_{F\,max\,2}| \le \delta_F, \tag{5.32}$$

where σ_{Fmax1} and σ_{Fmax2}: pinion and gear maximum tooth root tensile stresses; C_b: bending stress balance coefficient; δ_F: permissible stress balance tolerance (typically less than 2%–3%).

The root tensile stresses σ_{Fmax1} and σ_{Fmax2} are defined by the FEA. The bending stress balance coefficient C_b reflects a difference in the mating gear face widths, material properties, and the number of tooth load cycles of the pinion and the gear.

Direct Gear Design® uses the stress balance approach that utilizes an iteration method with the FEA stress calculation to satisfy the bending stress balance condition Equation (5.32) by adjusting the tooth thickness ratio (TTR), defined by Equation (5.27). The bending stress balance (Figure 5.54) works in combination with the tooth flanks and root fillet optimization.

---

[*] Calculations of the contact (flash) temperature and scuffing probability were done using the mesh scuffing risk analysis subroutine of the GearWin software package developed by Charles E. Long.

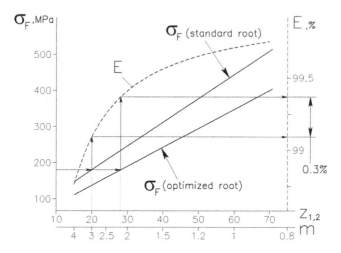

FIGURE 5.51
Conversion of potential bending stress reduction into increased gear mesh efficiency. (From Kapelevich, A.L., and Y.V. Shekhtman, Tooth fillet profile optimization for gears with symmetric and asymmetric teeth. *AGMA Fall Technical Meeting*, San Antonio, Texas, October 12–14, 2008, 08FTM06. With permission.)

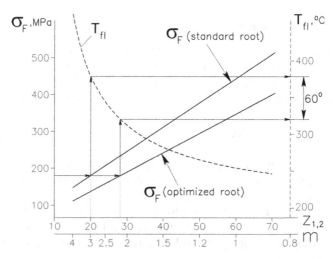

FIGURE 5.52
Conversion of potential bending stress reduction into contact (flash) temperature reduction.

5.2 Tooth Microgeometry

This section describes an optimization of the microgeometry that includes modification of the spur gear tooth flank profile to minimize transmission error variation and modification of the tooth line, aka lead crowning.

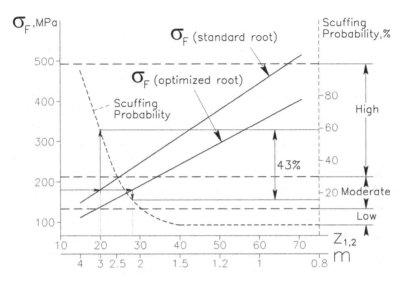

FIGURE 5.53
Conversion of potential bending stress reduction into scuffing probability reduction.

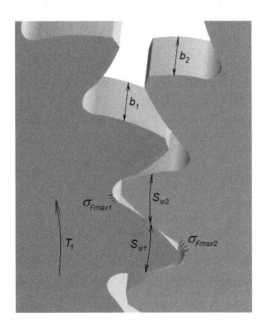

FIGURE 5.54
Tooth root stress balance; T_1: driving pinion torque, b_1 and b_2: pinion and gear face widths, S_{w1} and S_{w2}: pinion and gear tooth thicknesses at the operating pitch diameters, σ_{Fmax1} and σ_{Fmax2}: pinion and gear maximum tensile root stresses.

5.2.1 Tooth Flank Modification

The goal of the tooth flank microgeometry optimization [64] is to some degree compensate the influence of manufacturing tolerances, assembly misalignments, and deflections of gears (including teeth and body) and other gearbox components under operating loads, dynamic loads and inertia, thermal expansion or shrinkage, etc. All these factors distort a theoretically correct involute gear engagement by deviating the actual contact points from the ideal straight line of contact that increases transmission error variation and leads to increased noise and vibrations. The ultimate intention of the tooth flank microgeometry optimization is to make the real loaded gear pair operate as the ideal unloaded gear pair by modifying the tooth flank profile (Figure 5.55).

The tooth microgeometry optimization utilizes the same approach that was used in the effective contact ratio and transmission error calculation procedure (see Section 5.1.1.4), taking into account only the contact and bending gear tooth deflections. It is employed in combination with another iteration cycle that defines a tooth flank modification depth to achieve a minimal transmission error variation.

There are three most common tooth flank modification modes: a tip and root relief, an arc modification, and a parabolic crowning. Figure 5.56 shows the tooth flank modification modes as charts of the flank modification depth vs. the roll angle. The roll angle (see Figure 2.1) can be presented as follows:

$$\phi = \frac{180^{o}}{\pi}\tan\alpha, \tag{5.33}$$

where α is the involute profile angle.

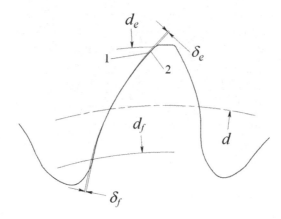

FIGURE 5.55
Tooth flank modification; 1: ideal involute profile, 2: modified flank profile; δ_f: flank modification depth at the form diameter d_f (near the tooth root), δ_e: flank modification depth at the effective tip diameter d_e (at the highest involute profile point near the tooth tip).

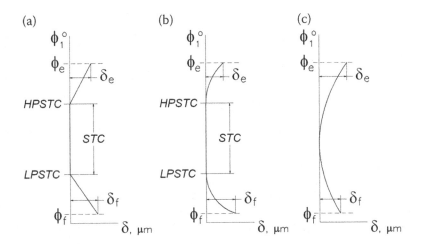

FIGURE 5.56
Tooth flank modification modes; (a): tip and root relief, (b): arc modification, (c): parabolic crowning; δ: flank modification depth in microns, δ_f: at the form diameter (near the tooth root), δ_e: at the effective tip diameter (at the highest involute profile point near the tooth tip); ϕ: roll angle in degrees, ϕ_f: at the form diameter, ϕ_e: at the effective tip diameter; LPSTC: lowest point of single tooth contact and HPSTC: highest point of single tooth contact for gears with a low medium contact ratio $\varepsilon_\alpha < 2.0$.

Although the tooth flank modification depth applied along the tooth flank length, it is commonly presented as a function of a roll angle because the involute inspection chart is developed by "rolling the gear on the base circle, producing contact traces of the profile" [35]. The tooth flank modification depth is a deviation from a straight line that represents the ideal flank. Figure 5.57 shows the tooth flank profile parabolic crowning modification charts as functions of a flank length (a) and a roll angle (b).

The relation between the length of the pinion tooth flank and roll angles is as follows:

$$L_1 = \frac{d_{b1}}{4}(\tan^2\alpha_{e1} - \tan^2\alpha_{f1}) = \frac{d_{b1}}{4}(\phi_{e1}^2 - \phi_{e1}^2), \qquad (5.34)$$

where α_{e1} and α_{f1}: involute profile angles at the effective tip diameter and the form diameter of the pinion; ϕ_{e1} and ϕ_{f1}: roll angles at the effective tip diameter and the form diameter of the pinion in radians.

The tip and root relief, and arc modification modes are applied to the low and medium contact ratio gears because they alter only the double tooth pair contact zones where a load sharing between two pairs of teeth occur. The parabolic crowning modifies a complete involute flank and can be applied for the low and medium contact ratio gears as well as the HCR gears. In a unidirectional asymmetric gear pair, a drive flank of the pinion is a subject for optimization. A coast flank of the pinion and both flanks of the driven gear

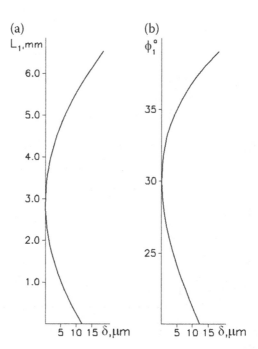

FIGURE 5.57
Flank modification depth; (a) along pinion flank, (b) as function of roll angle.

remain unmodified. However, when the optimization is done, a part of the modification depth can be transferred from a drive flank of the pinion to a drive flank of the mating gear. Most typical is a transfer of the pinion root relief depth to a tip relief depth of the mating gear (Figure 5.58). The modification depth δ_x at the pinion drive flank contact point X_1 with the profile angle α_{xd1} can be transferred to the mating gear drive flank contact point X_2. Then its profile angle α_{xd2} can be defined by altering Equations (2.106) and (2.110):

$$\tan \alpha_{xd1} \pm u \tan \alpha_{xd2} \mp (1 \pm u)\tan \alpha_{wd} = 0, \tag{5.35}$$

here and in Equations (5.36) and (5.37) the top sign (+ or −) is for the external gears and the bottom sign (+ or −) is for the internal gears.

Then the drive flank involute profile angles α_{xd1}, α_{xd2}, and α_{wd} should be replaced on the related roll angles from Equation (5.33)

$$\phi_{xd1} \pm u\phi_{xd2} \mp (1 \pm u)\phi_{wd} = 0 \tag{5.36}$$

and the mating gear drive flank contact point X_2 roll angle ϕ_{xd2} is

$$\phi_{xd2} = \frac{u \pm 1}{u}\phi_{wd} \mp \frac{1}{u}\phi_{xd1}. \tag{5.37}$$

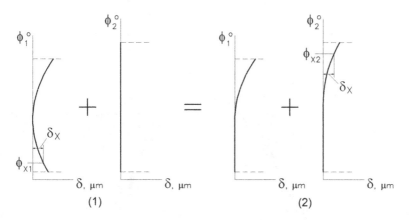

FIGURE 5.58
Flank modification depth transfer; (1): a gear pair with a completely optimized drive pinion flank, including the addendum and dedendum profile and an unmodified drive flank of the gear, (2): a gear pair with an optimized drive flank addendum of both the pinion and gear.

In some cases, like in an idler gear or a planet gear of an epicyclic gear stage, which has both drive flanks engaged with two different mating gears, the opposite flanks modifications should be optimized separately, considering differences in the flank engagements and loading. Besides, an asymmetric tooth of the idler gear or planet gear has different stiffness when loads are applied at the high and low pressure angle tooth flanks.

The following spur symmetric gear data and load condition are used to evaluate the tooth flank modification influence on the transmission error variation, effective contact ratio, and contact stress:

- numbers of pinion and gear teeth: 29;
- operating module: 3.0 mm;
- pressure angle values: from 18° to 40°;
- pinion and gear face widths: 25 mm;
- relative tooth tip thickness of pinion and gear: 0.30;
- gear material properties: Modulus of elasticity: 207,000 MPa, Poisson ratio: 0.3;
- all gears have optimized tooth root fillets;
- pinion torque: 400 Nm;
- tooth flank modification mode: parabolic crowning.

An overlay of the gear tooth profiles is shown in Figure 5.59.

Such gear pairs have a nominal contact ratio ranging ε_α = 0.91–2.36. An effective contact ratio under the load is ranging from ε_{ae} = 1.08–2.48. Charts of the nominal and effective contact ratio (before the flank modification)

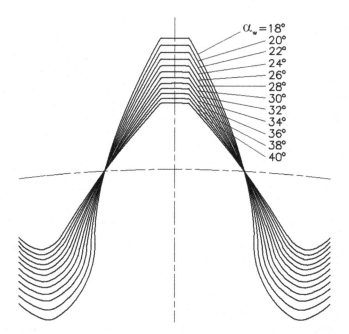

FIGURE 5.59
Gear tooth profile overlay.

are shown in Figure 5.60. These gear pairs are divided by three groups with a low-medium contact ratio (ε_α and $\varepsilon_{ae} < 2.0$), transitional contact ratio ($\varepsilon_\alpha < 2.0$, $\varepsilon_{ae} \geq 2.0$), and HCR ($\varepsilon_\alpha \geq 2.0$, $\varepsilon_{ae} > 2.0$).

Table 5.14 presents the parabolic crowning depths that are results of the tooth flank modification optimization to minimize the transmission error variation.

Transmission error variation of gear pairs with the unmodified pinion tooth flanks (line 1 in Figure 5.61) increases with the growing nominal contact ratio of the low and medium contact ratio gears reaching its maximum at the beginning of the transitional contact ratio gear zone. Then it drops sharply because of an increase of the tooth load sharing from one or two mating tooth pairs to two or three mating tooth pairs. Further increase of a nominal contact ratio of the HCR gears leads to the growth of the transmission error variation from its lowest value.

Parabolic crowning pinion tooth flank modification optimization (line 2 in Figure 5.61) allows for a significant reduction of the transmission error variation of the low-medium contact ratio (LCR) and HCR gear pairs. However, such pinion tooth flank modification optimization may not be advisable for the transitional contact ratio gears since transmission error variation is relatively small.

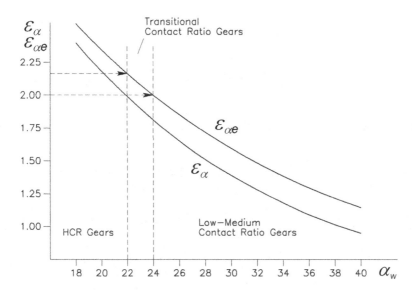

FIGURE 5.60

Nominal ε_α and effective $\varepsilon_{\alpha e}$ contact ratio charts.

TABLE 5.14

Parabolic Crowning Depth

Pressure angle,°	At effective tip diameter d_e, µm	At form diameter d_f, µm
18	31.0	13.8
20	13.7	11.6
22	3.5	5.0
4	21.1	27.7
26	17.7	18.7
28	14.5	12.6
30	11.5	8.3
32	9.0	5.3
34	6.8	3.1
36	5.0	1.7
38	3.2	0.7
40	1.6	0.2

Tooth flank modification affects the effective contact ratio. Figure 5.62 shows charts of the nominal contact ratio and effective contact ratio of gear pairs with the modified pinion flanks.

Tooth flank modification increases the curvature of the pinion flank profile affecting the contact stress (Figure 5.63). This change in the contact stress of the low-medium contact ratio (LCR) gears is negligible. However,

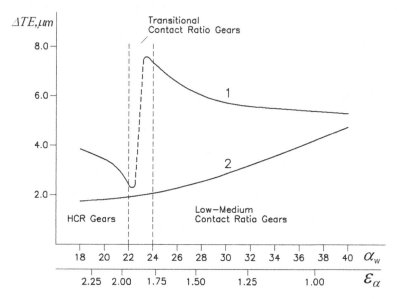

FIGURE 5.61
Transmission error variation before (1) and after (2) parabolic crowning pinion tooth flank modification optimization.

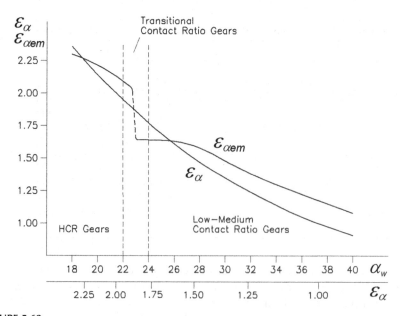

FIGURE 5.62
Nominal contact ratio ε_α and effective contact ratio of gear pairs with the modified pinion flanks $\varepsilon_{\alpha em}$.

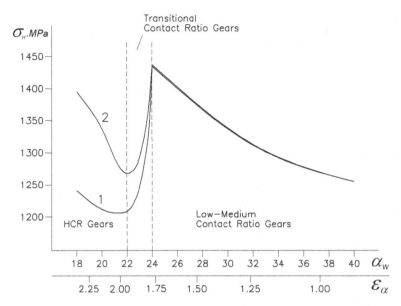

FIGURE 5.63
Contact stress charts before (1) and after (2) tooth flank modification.

the HCR gears may have the contact stress increased by 10% or even more as a result of the tooth flank modification. In this case, if the tooth flank modification compromises the tooth flank pitting or scuffing resistance beyond an acceptable level, it should not be applied.

Table 5.15 presents the LCR ratio asymmetric gear pair data and tooth flank optimization results, including maximum values of the tooth load, root and contact stresses, total tooth deflections, and transmission error variation. A total tooth deflection is a sum of the tooth bending and contact deflection. Three tooth flank modification modes (see Figure 5.56) are considered. A comparison of different modes of the drive flank modification optimization indicates an insignificant change in contact stress. The tip and root relief produces slightly greater transmission error variation reduction while keeping the relatively high effective drive contact ratio compared to the arc modification. The parabolic crowning also results in the minimal modification depth. Figure 5.64 presents the transmission error charts for the gear pairs with the initial unmodified drive flanks and gears with the optimized drive pinion flanks utilizing three different modification modes.

Figure 5.65 shows the pinion flank modification depth charts of for different modification modes.

Table 5.16 presents the asymmetric HCR gear pair data and tooth flank optimization results for the parabolic crowning modification mode, including maximum values of the tooth load, root and contact

TABLE 5.15

Low-Medium Contact Ratio (LCR) Gear Pair Tooth Flank Optimization

Gear	Pinion		Gear
Number of teeth	27		41
Module, mm		3.000	
Drive flank pressure angle		38°	
Coast flank pressure angle		19°	
Pitch diameter (PD), mm	81.000		123.000
Tooth tip diameter, mm	87.090		128.935
Root diameter, mm	74.393		116.230
Tooth thickness at PD, mm	4.807		4.618
Face width, mm	30.00		28.00
Center distance, mm		102.00	
Modulus of elasticity, MPa	207,000		207,000
Poisson ratio	0.3		0.3
Torque, Nm	700		1,063
Tooth load, N	21,936		21,936
Drive flank mesh efficiency		98.8% (average friction coefficient = 0.1)	
Nominal drive contact ratio		1.24	
Type of flank modification		No flank modification	
Effective drive contact ratio		1.46	
Root tensile stress, MPa	398		399
Contact stress, MPa		1,394	
Total tooth deflection, μm	15		20
Transmission error variation, μm		7.3	
Type of flank modification	Tip & root relief	Arc modification	Parabolic crowning
At effective tip diameter, μm	22	40	3
At form diameter, μm	6	11	12
Effective drive contact ratio	1.28	1.28	1.40
Root tensile stress, MPa	447/411*	454/416*	394/416*
Contact stress, MPa	1,398	1,400	1,400
Total tooth deflection, μm	18/22*	18/22*	15/21*
Transmission error variation, μm	3.3	3.7	3.7
Transmission error variation reduction	54%	49%	49%

Note:
* Pinion/gear

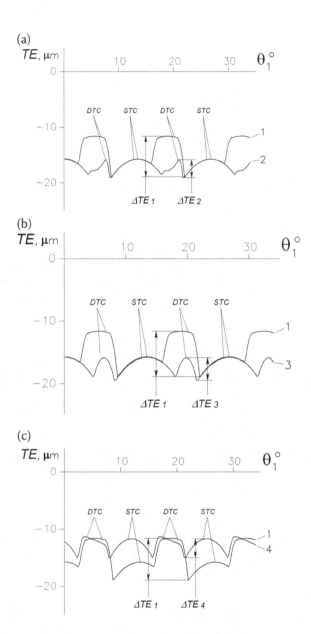

FIGURE 5.64
Transmission error charts and comparison of transmission error variations (ΔTE) of the gear pair 1 with unmodified pinion flanks to (a) the gear pair 2 with the pinion flank with the tip and root relief; (b) the gear pair 3 with the pinion flank with the arc modification; (c) the gear pair 4 with the pinion flank with the parabolic crowning; STC: single tooth pair contact, DTC: double tooth pair contact.

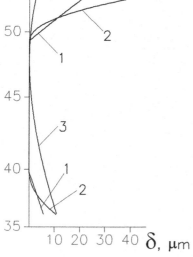

FIGURE 5.65
Pinion flank modification depth; 1: with the tip and root relief, 2: with the arc modification, 3: drive flanks of the pinion with the parabolic crowning; ϕ_1: pinion roll angle; δ: modification depth.

stresses, total tooth deflections, and transmission error variation. Figure 5.66 presents the transmission error charts for the gear pair with the initial unmodified drive flanks and gear pair with the optimized drive pinion flanks.

The pinion flank parabolic crowning modification depth chart of the HCR gear pair is shown in Figure 5.67.

Figures 5.68–5.72 show a comparison chart of the low-medium contact ratio (LCR) gear and high contact ratio (HCR) gear with the parabolic pinion flank modifications based on data from Tables 5.15 and 5.16. The HCR gears have several advantages over the low-medium contact ratio gears due to a much greater effective contact ratio and load sharing between two and three gear tooth pairs. Single tooth load and transmission error variation are decreased by about 60%. The root bending stress and contact stress reduction is not significant because the single tooth load decrease is practically nullified by the reduced tooth thickness at the root and lower pressure angle.

It is critical to understand that the microgeometry optimization defines the shape and depth of the drive flank modification for a particular transmitted torque value, for which the resulting modified flank profile provides minimal transmission error. For any other torque value, this flank profile is not optimal. For gear drives transmitting a constant torque, its value should be used for the driving flank modification optimization. However, there are many gear drives that operate at variable load values. In such cases, it is necessary to define which load condition is most damaging or critical for a specific gear drive application and use its value for the driving flank modification optimization.

TABLE 5.16

High Contact Ratio (HCR) Gear Pair Tooth Flank Optimization

Gear	Pinion		Gear
Number of teeth	27		41
Module, mm		3.000	
Drive flank pressure angle		24°	
Coast flank pressure angle		14°	
Pitch diameter (PD), mm	81.000		123.000
Tooth tip diameter, mm	89.443		131.542
Root diameter, mm	71.636		113.877
Tooth thickness at PD, mm	4.807		4.618
Face width, mm	30.00		28.00
Center distance, mm		102.00	
Modulus of elasticity, MPa	207,000		207,000
Poisson ratio	0.3		0.3
Torque, Nm	700		1,063
Tooth load, N	13,455		13,455
Root tensile stress, MPa	335		319
Drive flank mesh efficiency		98.0% (average friction coefficient = 0.1)	
Nominal drive contact ratio		2.04	
Type of flank modification		No flank modification	
Effective drive contact ratio		2.24	
Root tensile stress, MPa	341		314
Contact stress, MPa		1,260	
Total tooth deflection, μm	20		22
Transmission error, μm		4.4	
Type of flank modification		Parabolic crowning	
At effective tip diameter, μm		12	
At form diameter, μm		15	
Effective drive contact ratio		2.1	
Root tensile stress, MPa	364		332
Contact stress, MPa		1,362	
Total tooth deflection, μm	18		17
Transmission error variation, μm		2.1	
Transmission error variation reduction		52%	

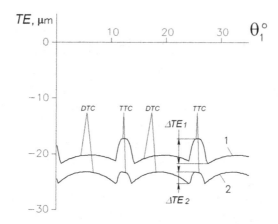

FIGURE 5.66
Transmission error charts and comparison of transmission error variations (*ΔTE*) of the gear pair 1 with unmodified pinion flanks with the gear pair 2 with the pinion flank with the parabolic crowning; STC: single tooth pair contact, DTC: double tooth pair contact.

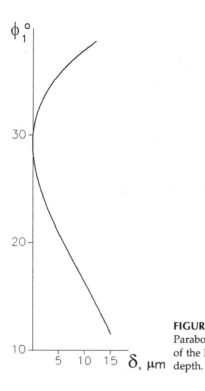

FIGURE 5.67
Parabolic crowning modification depth of the pinion flank of the HCR gear pair; ϕ_1: pinion roll angle; δ: modification depth.

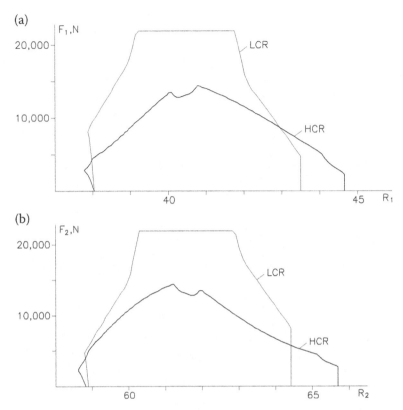

FIGURE 5.68
Single tooth load: (a) pinion, (b) gear; thin line for LCR gears, thick line for HCR gears.

5.2.2 Tooth Lead Crowning

Tooth lead crowning is used to compensate deviations from the theoretical tooth line, excluding the edge contact that typically results with a drastic increase in the contact and bending root stresses. These deviations are a combination of manufacturing and assembly tolerances and gear drive components deflections under operating conditions, including static and dynamic loads, temperature variation, etc. All these factors and their influence on tooth contact are described in [65]. There are different lead crowning modification modes, circular, parabolic, logarithmical, and others [66].

This section describes a simplified approach of the circular lead crowning modification depth definition based on the gear total tooth alignment tolerances (aka the total lead tolerance and total helix tolerance) and angular gear assembly misalignment.

Gear assembly misalignment angles (Figure 5.73) are defined by the gear supporting bearing center position tolerances Δ_{br1} and Δ_{br2}, and distance

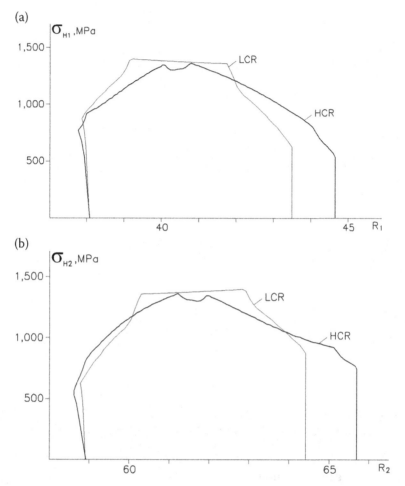

FIGURE 5.69
Contact stress: (a) pinion, (b) gear; thin line for LCR gears, thick line for HCR gears.

between bearings L. Maximum values of the parallel and skew misalignment angles are defined as follows:

$$\gamma_{p\,max} = \gamma_{s\,max} = \arctan\left(\frac{\Delta_{br1} + \Delta_{br2}}{L}\right). \tag{5.38}$$

A maximum value of the total misalignment angle (Figure 5.74) is

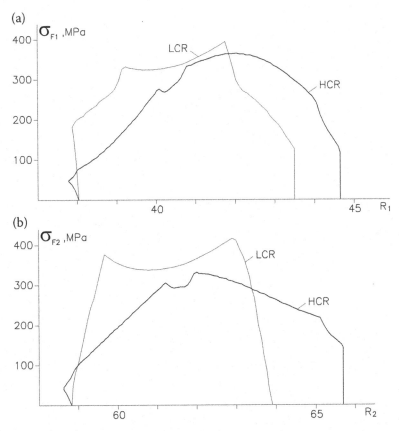

FIGURE 5.70
Root bending stress: (a) pinion, (b) gear; thin line for LCR gears, thick line for HCR gears.

$$\gamma_{max} = \gamma_{smax} + \arctan(\tan \gamma_{pmax} \times \tan \alpha_w) + \arctan\left(\frac{F_{\beta T1}}{b_w}\right) + \arctan\left(\frac{F_{\beta T2}}{b_w}\right),$$

(5.39)

where $F_{\beta T1}$ and $F_{\beta T2}$ – pinion and gear tooth alignment tolerances; b_w – effective face width.

Figure 5.75 shows a cross section of the contact of the pinion with the circular tooth lead crowning and the straight tooth line gear. The circular lead crowning radius R is

$$R = \frac{b_w}{2 \sin \gamma_{max}}.$$

(5.40)

The circular lead crowning depth Δ_{cr} is

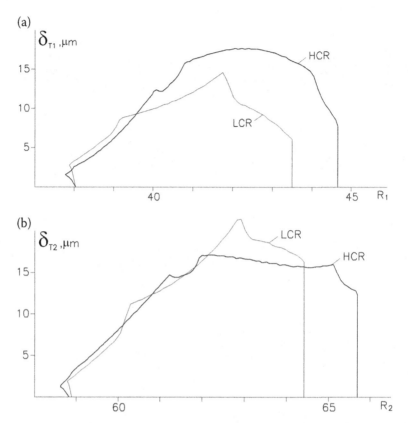

FIGURE 5.71
Total tooth deflection: (a) pinion, (b) gear; thin line for LCR gears, thick line for HCR gears.

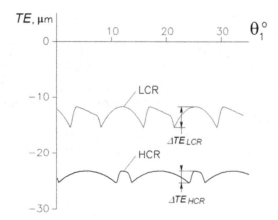

FIGURE 5.72
Transmission error variation: thin line for LCR gears, thick line for HCR gears.

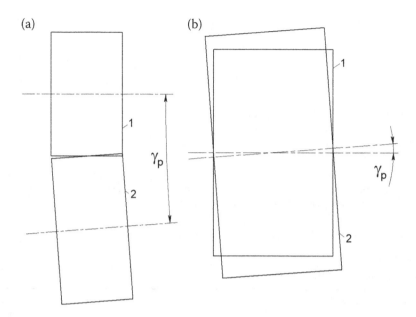

FIGURE 5.73
Gear assembly misalignment angles; (a) parallel misalignment angle γ_p, (b) skew misalignment angle γ_s; 1: pinion, 2: gear.

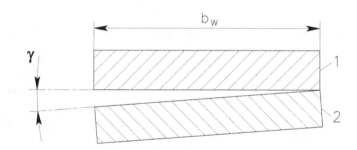

FIGURE 5.74
Cross section of the pinion 1 and gear 2 tooth contact; γ: total misalignment angle; b_w: effective face width.

$$\Delta_{cr} = R \times (1 - \cos \gamma_{max}). \tag{5.41}$$

Table 5.17 shows an example of the lead crowning depth calculation.

Typically, a crowning radius is very large, and a lead crowning depth is very small. This small deviation of the theoretical tooth line results in an insignificant increase in the contact stress.

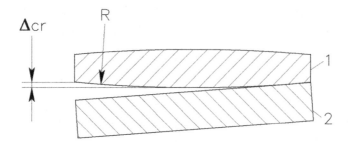

FIGURE 5.75
Cross section of the tooth contact of the pinion 1 with circular tooth lead crowning and gear 2;
R: lead crowning radius; Δ_{cr}: lead crowning depth.

TABLE 5.17

Lead Crowning Depth Calculation

Effective face width, mm	40.0
Pressure angle,°	25
Pinion and gear bearing center position tolerance, mm	0.030
Distance between bearings, mm	100.0
Max. Parallel and skew misalignment angles,	0.034
Total misalignment angle,°	0.065
Pinion crowning radius, mm	17,700
Pinion crowning depth, mm	0.011

Lead crowning is usually applied to the pinion flanks. However, the lead crowning depth can be shared between the pinion and gear flanks. An idler gear or planet gear of an epicyclic gear stage is engaged with two different mating gears. In this case, a lead crowning is applied to both flanks of the idler gear or planet gear, and the mating gears don't have a lead crowning.

6

Stress Analysis and Rating

Over the last 100 years, the maximum gear tooth root bending stress, characterizing tooth breakage, has been defined based on the Lewis equation. Gear design standards specify the dimensionless Lewis form factor Y as a function of the number of teeth, generating rack parameters, and its X-shift or addendum modification. The form factor Y, in combination with the stress correction factor K_f, accounting on stress concentration and load location, defines the bending strength geometry factor J.

The maximum gear contact stress, characterizing tooth surface pitting and wear breakage, is traditionally calculated by the Hertz equation. Gear design standards specify the pitting resistance geometry factor I accounting for the effects of radii of curvature, load sharing, and normal component of the transmitted load.

In some industries – for example, aerospace that is accustomed to using gears with nonstandard tooth shapes – the rating of these gears is established by comprehensive testing. However, such testing programs are not affordable for many other gear drive applications that could also benefit from a total gear tooth geometry optimization provided by Direct Gear Design®. Such optimized gears have involute flanks, the same as the conventional involute gears that are rated by national and international standards. This chapter presents the stress conversion coefficient method that allows using the existing gear standards to rate the optimized directly designed gears. Although it describes this approach to asymmetric gears, as a most general form of the directly designed involute gears, it is equally applicable to the symmetric tooth gears.

6.1 Stress Definition

The wide variability of parameters and shapes of directly designed gears makes it impossible to calculate their root bending stress by the Lewis equation. Before the introduction of the finite element analysis (FEA), stresses in the spur nonstandard gears were experimentally defined, utilizing the optical method of photoelasticity for measuring and visualizing stresses of the loaded tooth model, made out of transparent material with birefringence properties. Figure 6.1a shows a photoelastic model of the

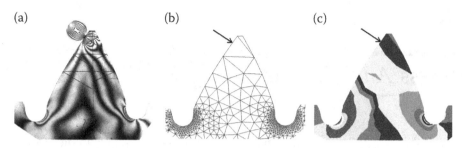

FIGURE 6.1
Asymmetric Tooth Stress Definition; (a) photoelastic modeling, (b) FEA mesh, (c) stress isograms.

asymmetric gear under load. Such gear stress definition was very time-consuming and costly. When the FEA became available, it significantly accelerated research and development of directly designed gears. Figures 6.1b and 6.1c present the FEA mesh and stress isograms.

For the tooth root fillet optimization (Section 5.1.2), the maximum root bending stress was preliminary defined by applying a 100% of the normal tooth load to the highest point of single tooth contact (HPSTC) for the low-medium contact ratio spur gears and about 2/3 of the normal tooth load at the highest point of double tooth contact for the high contact ratio spur gears. A similar approach is used for helical gears, but, in such a case, a stress analysis is done for the virtual spur gear pair that represents a cross-section of the helical gear pair that is normal to the tooth line.

After the tooth geometry of mating gears is finalized, the bending and contact stresses are defined considering the tooth pair load sharing tooth deflection under the applied load. In this case, the tooth load value is defined at different contact points of the involute tooth flank. Similar to the effective contact ratio and transmission error calculation procedure and tooth flank modification optimization, each angular position of the driven gear relative to the driving gear is iteratively defined by equalizing the sum of the tooth contact load moments of each gear to its applied torque. The related tooth contact loads are also iteratively defined to conform to tooth bending and contact deflections, where the tooth bending deflection in each contact point is determined based on the FEA-calculated flexibility and the tooth contact deflection is calculated by the Hertz equation.

Unlike the low number of the flank FEA nodes in preliminary stress calculation (Figure 6.1b), a final stress analysis utilizes a much greater number of the flank FEA nodes. Figures 6.2 and 6.3 show the FEA mesh and stress isograms for the external and internal tooth gears. The tooth load is consequently applied to each flank FE node from the lowest node near the form diameter to the highest one at the tooth tip for the driving pinion and from the highest node at the tooth tip to the lowest one near the form diameter for the driven gear. Each load value is defined considering load sharing with the neighboring teeth.

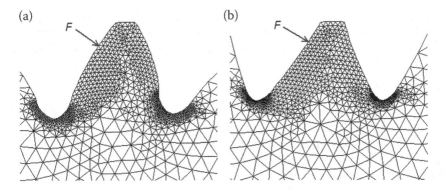

FIGURE 6.2
FEA mesh of the external tooth gear (a) and the internal tooth gear (b).

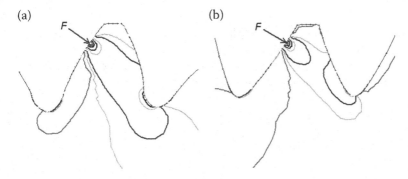

FIGURE 6.3
Stress isograms the external tooth gear (a) and the internal tooth gear (b).

The gear design and rating standards do not cover the directly designed gears. Their static root bending and contact stresses cannot be used for safety factor definition. It limits their broad implementation for various gear applications, despite potential performance advantages in comparison to traditionally designed gears. The following section presents an approach that makes it possible to apply the existing gear rating standards to nonstandard directly designed tooth gears.

6.2 Gear Rating

The suggested gear rating approach assumes using the comparable symmetric tooth gear pair, which can be evaluated with the existing gear rating standards [44,69]. This approach introduces the stress conversion coefficients, which establish relations between the directly designed gear pair and the comparable symmetric tooth gear pair geometries. These stress conversion coefficients allow for calculations of the directly designed gear pair stress safety factors.

6.2.1 Spur Gears

The comparable symmetric gear pair is traditionally defined for evaluation by the gear rating standards. The following equations utilize the directly designed gear pair parameters to calculate the generating gear rack parameters and addendum modifications or X-shifts of the comparable symmetric gear pair:

The generating rack module is

$$m = m_w = \frac{d_{w1}}{z_1} = \frac{d_{w2}}{z_2}, \tag{6.1}$$

where m_w is the operating module of the directly designed gear pair, $d_{w1,2}$ are the pitch diameters, $z_{1,2}$ are numbers of teeth.

The rack profile or pressure angle is an average value of the drive α_{wd} and coast α_{wc} pressure angles of the asymmetric tooth gear pair

$$a = (a_{wd} + a_{wc})/2. \tag{6.2}$$

For the symmetric tooth gear pair

$$a = a_w. \tag{6.3}$$

The rack addendum coefficient is defined by the tooth tip diameters $d_{a1,2}$ and pitch diameters $d_{w1,2}$ of directly designed gears:q

$$h_a = \frac{(d_{a1} - d_{w1}) \pm (d_{a2} - d_{w2})}{4m}, \tag{6.4}$$

where the "+" sign is for the external gears and the "−" sign is for the internal gears.

The gear rack should have a full tip radius to maximize the root strength of the comparable symmetric gears. The rack full tip radius coefficient is

$$r = \frac{\pi/4 - h_a \tan\alpha}{\cos\alpha}. \tag{6.5}$$

The radial clearance coefficient is

$$c = r \times (1 - \sin\alpha). \tag{6.6}$$

The rack dedendum coefficient is

$$h_d = h_a + c. \tag{6.7}$$

The addendum modification (X-shift) coefficients are

$$x_1 = \frac{S_{w1} - S_{w2}}{4m \times \tan\alpha} \quad and \quad x_2 = -x_1, \tag{6.8}$$

where $S_{w1,2}$ are the pinion and gear tooth thicknesses at the pitch diameters $d_{w1,2}$.

Figure 6.4 shows a schematic of generating of the comparable symmetric gear pair by the directly designed asymmetric gear pair parameters.

All other input data for the standard design of the comparable symmetric gear pair such as tolerances, operating parameters (loads, RPM, and lifetime), materials and heat treatment, lubrication, etc. should be the same as for the directly designed gear pairs for the directly designed gear pair. Table 6.1 presents an example of asymmetric and comparable symmetric external spur gear pair parameters.

Table 6.2 presents an example of asymmetric and comparable symmetric internal spur gear pair parameters.

The teeth of the spur asymmetric gears are shown in Figure 6.5.

The teeth of the comparable spur symmetric gears are shown in Figure 6.6.

The next step is a calculation of the bending and contact stress conversion coefficients that reflect differences in the directly designed and comparable symmetric gear geometry.

The bending stress conversion coefficients are defined as ratios of the maximum static tensile root stresses for pinion and gear of the comparable symmetric, and the analyzed directly designed (in this case, asymmetric) gear pairs

$$C_{F1,2} = \frac{\sigma_{F\,max(sym)1,2}}{\sigma_{F\,max(asym)1,2}}, \tag{6.9}$$

where $\sigma_{Fmax(sym)1,2}$ are the maximum static tensile root stresses for a pinion and gear of the comparable symmetric gear pair, and $\sigma_{Fmax(asym)1,2}$ are the maximum static tensile root stresses for a pinion and gear of the asymmetric gear pair.

These maximum static tensile root stresses are calculated by FEA under the same load.

The contact stress conversion coefficient is defined as a ratio of the maximum Hertz contact stresses of the comparable symmetric, and the analyzed directly designed (in this case, asymmetric) gear pairs

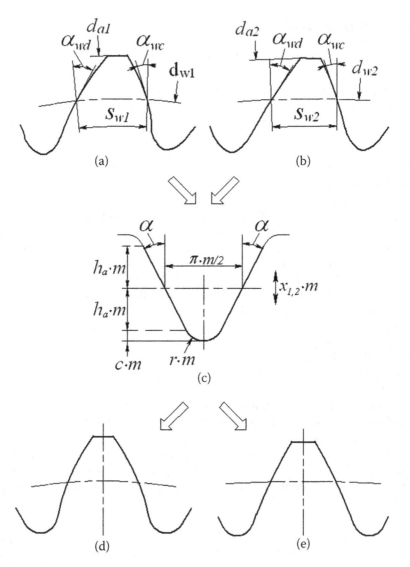

FIGURE 6.4
Schematic of generating of the comparable symmetric gear pair; (a) and (b) tooth parameters of the directly designed asymmetric pinion and gear, (c) generating gear rack parameters and X-shifts of the comparable symmetric gear pair, (d) and (e) comparable symmetric pinion and gear tooth profiles. (Kapelevich A.L. and Y.V. Shekhtman. Rating of asymmetric tooth gears, *AGMA Fall Technical Meeting*, Detroit, Michigan, October 18–20, 2015, 15FTM18. With permission.)

$$C_H = \frac{\sigma_{H\,max(sym)}}{\sigma_{H\,max(asym)}},\qquad (6.10)$$

TABLE 6.1

Asymmetric and Comparable Symmetric External Spur Gear Pair Parameters

Gear pair	Asymmetric		Comparable symmetric	
Design method	Direct Gear Design®		Rack generation	
Gear	Pinion	Gear	Pinion	Gear
Number of teeth	20	49	20	49
Module	5.000		5.000	
Pressure angle	35°/20°*		27.5°	
Asymmetry factor	1.147		1.0	
Pitch diameter (PD)	100.000	295.000	100.000	295.000
Base diameter	81.915/93.969*	200.692/230.225*	88.701	217.318
Tooth thickness at PD	8.168	7.540	8.168	7.540
Center distance	172.500		172.500	
Gear rack angle	–		27.5°	
Addendum coefficient	–		0.951	
Root radius coefficient	–		0.327	
Root clearance coefficient	–		0.176	
Dedendum coefficient	–		1.127	
X-shift coefficient	–	–	0.060	−0.060
Tip diameter	109.802	254.214	110.110	253.910
Root diameter	89.080	233.597	89.360	233.141
Root fillet profile	Optimized	Optimized	Trochoidal	Trochoidal
Face width	30.00	27.00	30.00	27.00
Contact ratio	1.20/1.55*		1.31	

Note
* Drive/coast flank

where $\sigma_{Hmax(sym)}$ is the maximum Hertz contact stress for a pinion and gear of the comparable symmetric gear pair, and $\sigma_{Hmax(asym)}$ is the maximum Hertz contact stress for pinion and gear of the asymmetric gear pair.

The rating of involute gears with symmetric teeth is established by the rating standards, for example, [44] or [69]. To apply these rating standards to the asymmetric tooth gears, the bending and contact safety factors that are defined for the comparable symmetric tooth gears by these standards should be multiplied by the contact and bending conversion coefficients accordingly. Then the bending safety factors of asymmetric tooth gears are as follows:

$$S_{F(asym)1,2} = C_{F1,2} \times S_{F(sym)1,2}, \qquad (6.11)$$

where $S_{F(sym)1,2}$ are the root bending safety factor of comparable symmetric tooth gears defined by a rating standard.

TABLE 6.2

Asymmetric and Comparable Symmetric Internal Spur Gear Pair Parameters

Gear pair	Asymmetric		Comparable symmetric	
Design method	Direct Gear Design®		Rack generation	
Gear	Pinion	Gear	Pinion	Gear
Number of teeth	20	49	20	49
Module	5.000		5.000	
Pressure angle	35**°/20°*		27.5°	
Asymmetry factor	1.147		1.0	
Pitch diameter (PD)	100.000	295.000	100.000	295.000
Base diameter	81.915/93.969*	200.692/230.225*	88.701	217.318
Tooth thickness at PD	8.168	7.540	8.168	7.540
Center distance	72.500		72.500	
Gear rack angle	–		27.5°	
Addendum coefficient	–		0.963	
Root radius coefficient	–		0.320	
Root clearance coefficient	–		0.172	
Dedendum coefficient	–		1.135	
X-shift coefficient	–	–	0.060	−0.060
Tip diameter	109.802	235.830	109.803	235.830
Root diameter	89.248	256.959	90.830	254.803
Root fillet profile	Optimized	Optimized	Trochoidal	Trochoidal
Face width	30.00	27.00	30.00	27.00
Contact ratio	1.24/1.83*		1.44	

Note
* Drive/coast flank

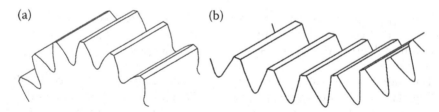

FIGURE 6.5
Spur asymmetric teeth; (a) external gear, (b) internal gear.

The contact safety factor of asymmetric tooth gears is

$$S_{H(asym)} = C_H \times S_{H(sym)}, \tag{6.12}$$

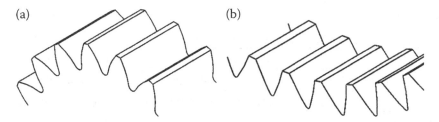

FIGURE 6.6
Comparable spur symmetric teeth; (a) external gear, (b) internal gear.

TABLE 6.3

Asymmetric and Comparable Symmetric External Spur Gear Pair Stress Analysis

Gear pair	Asymmetric		Comparable symmetric	
Design method	Direct Gear Design®		Rack generation	
Gear	Pinion	Gear	Pinion	Gear
Number of teeth	20	49	20	49
Module	5.000		5.000	
Pressure angle	35°/20°*		27.5°	
Face width	30.00	27.00	30.00	27.00
Torque, Nm	900	2205	900	2205
RPM	1000	408	1000	408
Service life, hrs	2000		2000	
Material type	Carburized, case hardened steel, AISI 8620			
Bending stress, MPa	–	–	448 b	480**
Bending stress (FEA), MPa	276	277	309	334
Bending stress conversion coefficients (FEA)	1.120	1.206	–	–
Contact stress, MPa	–	–	1507**	1407**
Contact stress (Hertz), MPa	1257		1349	
Contact stress conversion coefficient (Hertz)	1.073		–	
Bending safety factors	1.90	1.95	1.70**	1.62**
Contact safety factors	1.02	1.12	0.95**	1.04**

Notes
 * Drive/coast flank
** Calculation Method: per ISO 6336 standard

TABLE 6.4

Asymmetric and Comparable Symmetric Internal Spur Gear Pair Stress Analysis

Gear pair	Asymmetric		Comparable symmetric	
Design method	Direct Gear Design®		Rack generation	
Gear	Pinion	Gear	Pinion	Gear
Number of teeth	20	49	20	49
Module	5.000		5.000	
Pressure angle	35°/20°*		27.5°	
Face width	30.00	27.00	30.00	27.00
Torque, Nm	900	2205	900	2205
RPM	1000	408	1000	408
Service life, hrs	2000		2000	
Material type	Carburized, case hardened steel, AISI 8620			
Bending stress, MPa	–	–	417**	388**
Bending stress (FEA), MPa	270	239	292	322
Bending stress conversion coefficients (FEA)	1.08	1.35	–	–
Contact stress, MPa	–	–	1026**	889**
Contact stress (Hertz), MPa	780		835	
Contact stress conversion coefficient (Hertz)	1.070		–	
Bending safety factors	2.30	3.17	2.13**	2.35**
Contact safety factors	1.53	1.81	1.43**	1.69**

Notes
* Drive/coast flank
** Calculation Method: per ISO 6336 standard

where $S_{H(sym)}$ is the flank contact (pitting) safety factor of comparable symmetric tooth gears defined by a rating standard.

Examples of the spur asymmetric and comparable symmetric external and internal spur gear stress analysis results are presented in Tables 6.3 and 6.4.

6.2.2 Helical Gears

Rating of helical directly designed gears [68] by the existing rating standards employs the same approach as for the spur gears. The standard design of helical symmetric gears is based on parameters of the virtual spur gears that represent normal sections of the helical gears. Accordingly, all parameters of the generating rack of comparable helical symmetric gears can be defined for the virtual spur gears in the normal section.

The comparable symmetric gear helix angle is

$$\beta = \beta_w, \tag{6.13}$$

where β_w is the operating helix angle of the directly designed gear pair.

The normal module of the generating rack is

$$m_n = m_w \times cos\beta_w = \frac{d_{w1}}{z_1} \times cos\beta_w = \frac{d_{w2}}{z_2} \times cos\beta_w, \tag{6.14}$$

The generating rack profile or pressure angle is

$$\alpha_n = acos((cos\alpha_{wd} + cos\alpha_{wc})/2) \times tan\beta_w. \tag{6.15}$$

For the symmetric tooth gears

$$\alpha_n = acos(cos\alpha_w) \times tan\beta_w. \tag{6.16}$$

The rack addendum coefficient is

$$h_a = \frac{(d_{a1} - d_{w1}) \pm (d_{a2} - d_{w2})}{4m_n}, \tag{6.17}$$

where the "+" sign is for the external gears and the "−" sign is for the internal gears.

The gear rack should have a full tip radius to maximize the root strength of the comparable symmetric gears. The rack full tip radius coefficient is

$$r = \frac{\pi/4 - h_a tan\alpha}{cos\alpha}. \tag{6.18}$$

The radial clearance coefficient is

$$c = r \times (1 - sin\alpha). \tag{6.19}$$

The rack dedendum coefficient is

$$h_d = h_a + c. \tag{6.20}$$

The addendum modification (X-shift) coefficients are

$$x_1 = \frac{(s_{w1} - s_{w2}) \times cos\beta}{4m_n \times tan\alpha_n} \quad and \quad x_2 = -x_1. \tag{6.21}$$

Figure 6.7 shows a schematic of generating of the comparable helical symmetric gear pair by the directly designed helical asymmetric gear pair parameters.

The root bending and contact stress conversion coefficients, and safety factors of the helical gears are defined by Equations (6.9–6.12).

Tables 6.5 and 6.6 present examples of the helical asymmetric and comparable symmetric external and internal gear pair parameters.

The teeth of the helical asymmetric gears are shown in Figures 6.8 and 6.9.

The teeth of the comparable helical symmetric gears are shown in Figure 6.9.

TABLE 6.5

Asymmetric and Comparable Symmetric External Helical Gear Pair Parameters

Gear pair	Asymmetric		Comparable symmetric	
Design method	Direct Gear Design®		Rack generation	
Number of teeth	17	23	17	23
Normal module	4.000		4.000	
Normal pressure angle	35°/18°*		26.5°	
Asymmetry factor	1.179		1.0	
Helix angle	20°		20°	
Lead of helix	624.607	845.057	624.607	845.057
Helix hand	Right	Left	Right	Left
Pitch diameter (PD)	72.364	97.904	72.364	97.904
Base diameter	58.026/68.39*	78.506/92.529*	63.924	86.485
Normal tooth thickness at PD	6.390	6.176	6.390	6.176
Center distance	85.134		85.134	
Normal gear rack angle	–		26.5°	
Addendum coefficient	–		1.051	
Root radius coefficient	–		0.292	
Root clearance coefficient	–		0.162	
Dedendum coefficient	–		1.213	
X-shift coefficient	–	–	0.025	–0.025
Tip diameter	80.845	106.236	80.985	106.100
Root diameter	62.967	88.318	62.873	87.988
Root fillet profile	Optimized	Optimized	Trochoidal	Trochoidal
Face width	40.00	37.00	40.00	37.00
Transverse contact ratio	1.20/1.54*		1.32	
Axial contact ratio	1.01		1.01	
Total contact ratio	2.21/2.55		2.33	

Note
* Drive/coast flank

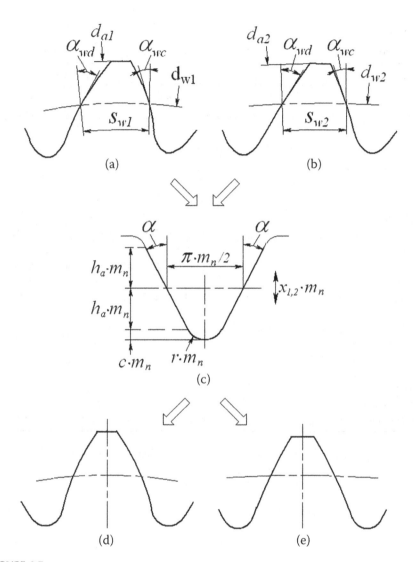

FIGURE 6.7
Schematic of generating of the comparable helical symmetric gear pair; (a) and (b) tooth parameters of the directly designed asymmetric pinion and gear in the transverse section, (c) generating gear rack parameters and X-shifts of the comparable symmetric gear pair in normal section, (d) and (e) comparable symmetric pinion and gear tooth profiles in the transverse section. (Kapelevich A.L. and Y.V. Shekhtman. Rating of helical asymmetric tooth gears, *Gear Technology*. November/December 2017, 78–81. With permission.)

Examples of the helical asymmetric and comparable symmetric external and internal spur gear stress analysis results are presented in Tables 6.7 and 6.8.

It is important to understand that the described comparable symmetric gear pairs represent the virtual gears that are designed based on the directly

TABLE 6.6

Asymmetric and Comparable Symmetric External Helical Gear Pair Stress Analysis

Gear pair	Asymmetric		Comparable symmetric	
Design method	Direct Gear Design®		Rack generation	
Gear	Pinion	Gear	Pinion	Gear
Number of teeth	17	23	17	23
Module	4.000		4.000	
Normal pressure angle	35°/18°*		26.5°	
Helix angle	20°		20°	
Face width	40.00	37.00	40.00	37.00
Torque, Nm	700	947	700	947
RPM	1000	739	1000	739
Service life, hrs	2000		2000	
Material type	Carburized, case hardened steel, AISI 8620			
Bending stress, MPa	–	–	369**	392**
Bending stress (FEA), MPa	271	285	298	315
Bending stress conversion coefficients (FEA)	1.100	1.105	–	–
Contact stress, MPa	–	–	1485**	1485**
Contact stress (Hertz), MPa	1282		1340	
Contact stress conversion coefficients (Hertz)	1.045		–	
Bending safety factors	2.64	2.51	2.40**	2.27**
Contact safety factors	1.06	1.07	1.01**	1.02**

Notes
 * Drive/coast flank
 ** Calculation Method: per ISO 6336 standard

(a) (b)

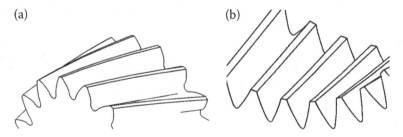

FIGURE 6.8
Helical asymmetric teeth; (a) external gear, (b) internal gear.

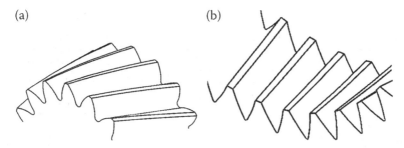

FIGURE 6.9
Comparable helical symmetric teeth; (a) external gear, (b) internal gear.

TABLE 6.7

Asymmetric and Comparable Symmetric Internal Helical Gear Pair Parameters

Gear pair	Asymmetric		Comparable symmetric	
Design method	Direct Gear Design®		Rack generation	
Number of teeth	17	41	17	41
Normal module	4.000		4.000	
Normal pressure angle	35°/18°*		26.5°	
Asymmetry factor	1.179		1.0	
Helix angle	20°		20°	
Lead of helix	624.607	1506.405	624.607	845.057
Helix hand	Right	Left	Right	Left
Pitch diameter (PD)	72.364	174.525	72.364	174.525
Base diameter	58.026/68.391*	139.946/ 164.943*	63.923	154.264
Normal tooth thickness at PD	6.390	6.176	6.390	6.176
Center distance	51.080		51.080	
Normal gear rack angle	–		26.5°	
Addendum coefficient	–		1.026	
Root radius coefficient	–		0.306	
Root clearance coefficient	–		0.170	
Dedendum coefficient	–		1.196	
X-shift coefficient	–	–	0.027	−0.027
Tip diameter	80.838	166.343	80.838	166.343
Root diameter	62.788	183.780	64.182	182.999
Root fillet profile	Optimized	Optimized	Trochoidal	Trochoidal
Face width	40.00	37.00	40.00	37.00
Transverse contact ratio	1.25/1.69*		1.71	
Axial contact ratio	1.01		1.01	
Total contact ratio	2.26/2.70		2.72	

Note
* drive/coast flank

TABLE 6.8

Asymmetric and Comparable Symmetric Internal Helical Gear Pair Stress Analysis

Gear pair	Asymmetric		Comparable symmetric	
Design method	Direct Gear Design®		Rack generation	
Gear	Pinion	Gear	Pinion	Gear
Number of teeth	17	41	17	41
Module	4.000		4.000	
Normal pressure angle	35°/18°*		26.5°	
Helix angle	20°		20°	
Face width	40.00	37.00	40.00	37.00
Torque, Nm	700	1688	700	1688
RPM	1000	415	1000	415
Service life, hrs	2000		2000	
Material type	Carburized, case hardened steel, AISI 8620			
Bending stress, MPa	–	–	362**	320**
Bending stress (FEA), MPa	284	246	298	314
Bending stress conversion coefficients (FEA)	1.05	1.276	–	–
Contact stress, MPa	–	–	843**	843**
Contact stress (Hertz), MPa	783		840	
Contact stress conversion coefficients (Hertz)	1.073		–	
Bending safety factors	2.57	3.63	2.45**	2.85**
Contact safety factors	1.83	1.89	1.71**	1.76**

Notes
 * Drive/coast flank
 ** Calculation Method: per ISO 6336 standard

designed gear parameters with a sole purpose to rate these nonstandard gear pairs utilizing the existing rating standards for symmetric involute gears. Differences in safety factors of these nonstandard and comparable symmetric gear pairs must not be used to conclude potential load capacity and performance of directly designed gear pairs. For this purpose, the directly designed gears should be compared to the existing benchmark traditionally designed symmetric gears that are considered as the best gear solutions for a particular application.

7

Plastic and Powder Metal Gear Design Specifics

Plastic injection molding and powder metal (PM) processing are highly cost-effective gear fabrication technologies for moderately to highly loaded mass-produced gear drives. Traditional metal gear design is constrained by preselected, typically standard, tooling rack proportions and gear generation machining methods. Plastic injection molding and PM gear processing do not utilize a rack generation, and the design of these gears should not be affected by the machined gear constraints. This allows for complete optimization of gear tooth geometry to maximize the gear drive performance for a particular application. The chapter describes the specifics of the Direct Gear Design® of plastic and PM gears.

7.1 Plastic Gears

Nowadays, progress in polymer materials and injection molding processing has allowed a drastic expansion of plastic gear applications. They are used not only for lightly loaded motion transmissions, but also in moderate load power drives in automotive, agriculture, medical, robotics, and many other industries.

There are comprehensive books about plastic gears written by C. Adams [70] and V.E. Starzhinsky with coauthors [71]. The AGMA standards [72,73] present tooth proportions for plastic gears, and the AGMA standard [74] gives a polymer material selection guideline. The VDI 2736 standard [75] describes a material selection, design and tolerancing, and production methods of thermoplastic gears. The paper [76] describes polymer gear wear behavior and its performance prediction based on extensive investigations on thermal-mechanical tooth contact both numerically and experimentally.

Benefits of polymer gears in comparison to metal ones include:

- low-cost injection molding processing;
- low vibration and noise;
- low weight and inertia;

- low or no corrosion;
- no electric current conductivity;
- in some applications, plastic gears can work without external (oil or grease) lubrication.

These advantages made possible usage of polymers for a wide variety of gear drives. However, some limitations must be taken into account considering plastic gears:

- low strength and wear resistance;
- low thermal conductivity and operating temperature;
- a significant deviation of material property parameters;
- sensitivity to operating conditions (temperature and humidity);
- low modulus of elasticity and increased tooth deflection;
- limited injection molding process accuracy;
- creep.

The main polymer gear materials are acetals (POM) and nylons, polyesters, and polycarbonates. They can be used with operating temperatures up to 150°C. For elevated temperature (<170°C) suitable gear polymers are polyphthalamide (PPA), nylon 46, and similar. High-temperature (<200°C) plastic materials include polyetherimide (PEI), polyetheretherketone (PEEK), and liquid crystal polymers (LCPs).

Different additives to a polymer composition improve some drawbacks of gear plastics properties. Additives for higher flexural strength include glass, carbon, and aramid (Kevlar) fibers. Tooth flank wear resistance of non-lubricated plastic gears can be increased by antiwear and antifriction additives: silicone, polytetrafluoroethylene (PTFE), graphite powders, molybdenum disulfide (MoS_2), etc.

Dissimilar polymers should be used for nonlubricated mating gears to avoid squeaking noise.

Table 7.1 shows an example of a comparison of the traditionally and directly designed plastic gear pairs with identical numbers of teeth, center distance, tooth thicknesses, and face widths. Gear materials are also the same. The traditionally designed gear pair has the standard tooth proportions. The gear pair defined by Direct Gear Design® has a significantly larger pressure angle and optimized tooth root fillets. Its pinion tooth flanks have an optimized parabolic modification. These design features allow achieving a significant reduction in the root and contact stresses and also in transmission error variation compared to the traditionally designed gear pair without increasing fabrication cost.

Besides original plastic gear designs, polymer gears are often considered for a replacement of already existing relatively lightly loaded metal gears,

TABLE 7.1

Traditional *vs* Direct Plastic Gear Design

Design method		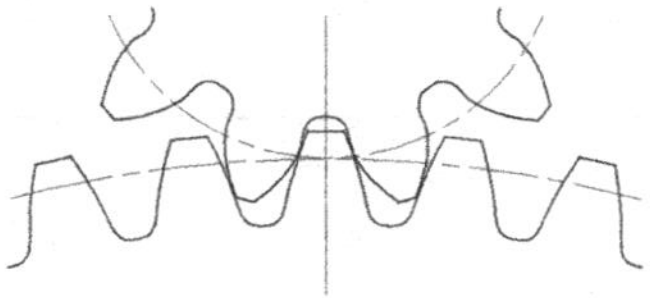 Traditional Gear Design		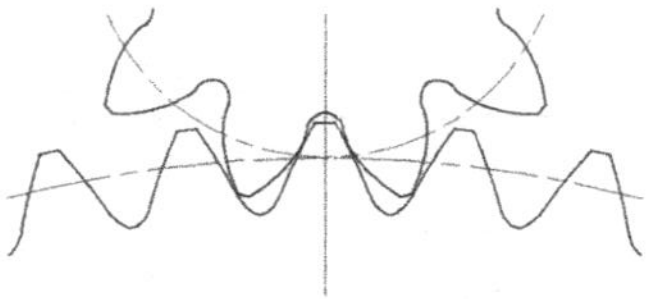 Direct Gear Design®	
Gear		Driving	Driven	Driving	Driven
Number of teeth		11	57	11	57
Normal module, mm		1.250		1.250	
Pressure angle		20°		27°	
Generating rack coefficients	Addendum	1.0		N/A	N/A
	Dedendum	1.25		N/A	N/A
	Tip radius	0.3		N/A	N/A
	Radial clearance	0.25		N/A	N/A
X-shift coefficient		0.3	−0.3	N/A	N/A
Tooth tip thickness coefficient		0.43	0.82	0.25	0.25
Pitch diameter (PD), mm		13.750	71.250	13.750	71.250
Base diameter, mm		12.921	66.953	12.251	63.484
Tooth tip diameter, mm		16.887	72.859	16.677	73.696
Root diameter, mm		11.292	67.313	11.061	67.958
Root fillet		Trochoidal	Trochoidal	Optimized	Optimized
Tooth thickness at PD, mm		2.206	1.668	2.206	1.668
Normal backlash, mm		0.050		0.050	

(*Continued*)

TABLE 7.1 (Continued)

Design method

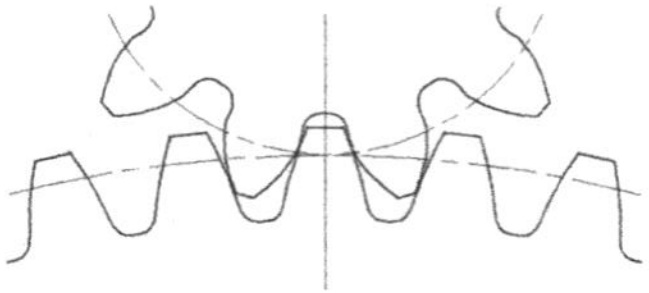

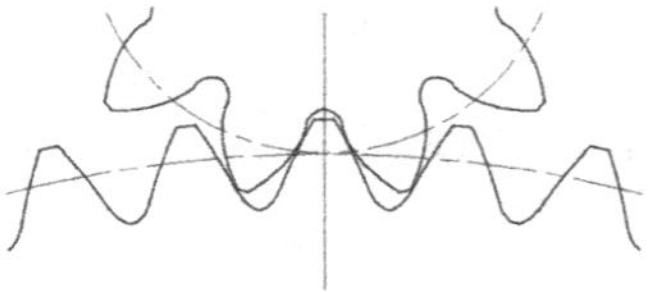

Gear	Traditional Gear Design		Direct Gear Design®	
	Driving	Driven	Driving	Driven
Center distance, mm	42.500		42.500	
Face width, mm	12.0	11.0	12.0	11.0
Nominal contact ratio	1.41		1.44	
Mesh efficiency, %	97.5		97.6	
Maximum driving torque, Nm	1.5		1.5	
Gear material	Delrin® 100	Zytel® 101L[*]	Delrin® 100	Zytel® 101L[*]
Flexural modulus, MPa	2,900	1,200	2,900	1,200
Poisson ratio	0.37	0.43	0.37	0.43
Effective contact ratio	1.88		1.82	
Yield tensile strength, MPa	72	55	72	55
Root stress (FEA), MPa	41.8	48.7	32.7(−22%)	31.8 (−35%)
Contact stress (Hertz), MPa	81.5		57.6 (−29.4%)	
Transmission error variation	56.8		33.2 (−41.5%)[**]	

Notes
[*] At 50% relative humidity
[**] Optimized parabolic flank modification

usually for cost and noise reduction. In such cases, the exact replication of a metal gear design typically does not work, mostly because of the low strength of polymers in comparison to metals. Table 7.2 presents an example of such a metal-to-plastic gear conversion.

TABLE 7.2

Metal-to-Plastic Gear Conversion

Design method		Metal gear pair Traditional Gear Design		Plastic gear pair Direct Gear Design®	
Gear		Driving	Driven	Driving	Driven
Number of teeth		32	56	24	42
Normal module, mm		1. 500		2.000	
Pressure angle		20°		26°	
Generating	Addendum	1.0		N/A	
Rack	Dedendum	1.25		N/A	
Coefficients	Tip radius	0.3		N/A	
	Radial clearance	0.25		N/A	
X-shift coefficient		0.0	0.0	N/A	N/A
Tooth tip thickness coefficient		0.73	0.76	0.25	0.25
Pitch diameter (PD), mm		48.000	84.000	48.000	84.000
Base diameter, mm		45.105	78.934	43.142	75.499
Tooth tip diameter, mm		50.924	72.427	53.195	88.163
Root diameter, mm		44.162	86.925	43.468	78.344
Root fillet		Trochoidal	Trochoidal	Optimized	Optimized
Tooth thickness at PD, mm		2.330	2.330	3.487	2.740
Normal backlash, mm		0.050		0.050	
Center distance, mm		66.000		66.000	
Face width, mm		13.0	12.0	13.0	12.0
Nominal contact ratio		1.66		1.65	
Mesh efficiency, %		98.7		98.3	
Operating temperature, °C		80		80	
Maximum driving torque, Nm		10.0	–	10.0	–
Gear material		Steel AISI-1144, annealed		Victrex HPG™ 140 GRA	
Flexural modulus, MPa		200,000		3,500	
Poisson ratio		0.29		0.4	
Effective contact ratio		1.76		2.08	
Yield tensile strength, MPa		345		70	
Root stress (FEA), MPa		56.6	61.6	27.8	27.8
Root safety factor		6.1:1	5.6:1	2.5:1	2.5:1
Compressive strength, MPa		794		100	
Contact stress (Hertz), MPa		504		54.4	
Flank safety factor		1.6:1		1.8:1	

The plastic gear pair defined by Direct Gear Design® has reduced numbers of teeth and increased modules to keep the same center distance and gear ratio as the metal machined gears. It also has a larger pressure angle and effective contact ratio and optimized tooth root fillets. All these alterations make it possible to replace the machined metal gears with the plastic molded gears providing sufficient safety factors. This replacement delivers not only required load carrying capacity but also significantly reduces gear production cost, and noise and vibration [77].

Overlays of the metal and symmetric plastic gear tooth profiles are shown in Figure 7.1.

The following guidelines should be considered for high performance plastic gear design:

- increasing safety factors to guarantee sufficient tooth strength considering a wide range of material properties deviation and dependence on molding process parameters, operating temperature, humidity, etc.;

- increasing tooth size (larger module or coarser diametral pitch) to reduce root bending stress and reducing numbers of teeth to keep the required gear ratio and center distance (this also reduces tolerance sensitivity);

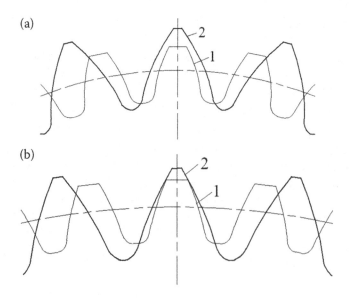

FIGURE 7.1
Metal (1) and plastic (2) gear tooth profile comparison; (a) pinion teeth, (b) gear teeth. (From Kapelevich A.L. Optimal polymer gear design: metal-to-plastic conversion, *AGMA Fall Technical Meeting*, Detroit, Michigan, October 14–16, 2019, 19FTM04. With permission.)

- optimizing tooth flanks to achieve an effective (under load) contact ratio ≥ 2.0 and at the same time higher operating pressure angle, which is possible considering the low flexural module of polymers compared to steels (this allows for the distribution of the trans-mitted load between at least two tooth pairs, significantly reducing tooth flank and root stresses and increasing tooth flank wear resistance);
- optimizing the root fillet profile for root bending stress reduction;
- bending stress balance that equalizes safety factors of mating gears;
- specific sliding balance for higher efficiency and tooth flank wear resistance;
- applying an asymmetric tooth profile for unidirectionally loaded gear drives.

7.2 Powder Metal Gears (Written by Dr. Anders Flodin)

The primary reason that powder metal gears may be preferred over more traditional gear materials is fabrication cost. In large production quantities, powder metal gears are less expensive compared to steel ones. Other benefits include reduced weight, inertia, and noise and vibration. The Direct Gear Design® method is particularly suitable for PM gears because any tooth profile, including the root shape, can be implemented in the compaction tooling. Any tooth line barrel crowning, which might be im-possible to compact, can be achieved by the rolling densification process (see Figure 9.22 in Section 9.1.2.1) in the sintered state or hard machining in the hardened state. The grinding protuberance used for PM gears will be the same as for wrought steel, normally 0.1 mm for PM relevant sized gears. The tolerances after the sintering of a PM gear are usually better than after hobbing a wrought gear [78], so there is room to reduce the protuberance to speed up a power honing process. Achieving the required protuberance of the low pressure angle coast flank of asymmetric gears by hobbing can be difficult. This problem does not exist with PM gear com-pacting, which allows for an increased difference in the pressure angle of the drive and coast flanks of the tooth. The gear teeth could also be compacted tapered, like in the reverse gear of the Smart Fortwo trans-mission (Figure 7.2).

The HCR gears used for noise and vibration reduction have slender teeth, which in combination with the optimized root fillet, make them difficult to machine by hobbing. However, this tooth shape is perfectly suitable for PM processing. Besides, reduction in Modulus elasticity and Poisson ratio

FIGURE 7.2
Tapered teeth of the reverse gear of the Smart Fortwo transmission. (Courtesy of Höganäs AB, Höganäs, Sweden. With permission.)

(Equations 5.30 and 5.31), due to lower PM alloy density compared to solid steel, results in an additional decrease of noise and vibration.

Table 7.3 presents a comparison of the PM gears designed traditionally (FZG gear data) and by the Direct Gear Design® method with two optimized geometry options, with symmetric and asymmetric tooth profiles. The directly designed gears have the same main parameters and dimensions, but their root fillet profiles are optimized. They also have an optimized parabolic modification of the driving gear tooth flanks. Tooth profiles of the comparable gear pairs are shown in Figure 7.4.

In comparison to the traditionally designed baseline gear pair, the optimized directly designed gears provide a substantial reduction in root stresses and transmission error variation that decreases gear noise and vibration. The asymmetric tooth gears, although having about the same root tension stresses as the optimized symmetric tooth gears, provide a significant reduction in contact stress and an additional decrease of transmission error variation. Their root compression stresses are relatively high, just slightly lower than that for the baseline gear pair.

Powder metal gears are usually lower in weight than the wrought counterparts unless the PM gears are powder forged. Single compaction high-density PM gears typically have a density of around 7.2g/cc. It will give an 8% weight reduction by just switching material from solid steel to PM. But the production process lends itself well to weight reduced gear bodies with thinner web sections and holes without producing any waste material [79]. The designer should make use of this since it not only creates

TABLE 7.3

Traditional vs Direct Gear Design® of Powder Metal Symmetric and Asymmetric Gears

Design method Tooth profile	Traditional Symmetric, baseline (Figure 7.3a)		Direct Gear Design®			
			Symmetric, optimized (Figure 7.3b)		Asymmetric, optimized (Figure 7.3c)	
Gear	Driving	Driven	Driving	Driven	Driving	Driven
Number of teeth	16	24	16	24	16	24
Nominal module	4.50	4.50	4.575	4.575	4.575	4.575
Operating module	4.575	4.575	4.575	4.575	4.575	4.575
Nominal pressure angle	20°	20°	22.44°	22.44°	36°/20°*	36°/20°*
X-shift coefficient	+1.82	+1.71	N/A	N/A	N/A	N/A
Operating pitch diameter (OPD)	73.2	109.8	73.2	109.8	73.2	109.8
Base diameter	67.658	101.487	67.658	101.487	59.220/ 68.786*	88.830/ 103.178*
Outer diameter	82.46	118.36	82.46	118.36	82.46	118.36
Root diameter	62.39	98.29	63.41	99.01	63.50	99.67
Tooth thickness at OPD	7.321	7.051	7.321	7.051	7.321	7.051
Face width	14.0	14.0	14.0	14.0	14.0	14.0
Center distance	91.500		91.500		91.500	
Operating pressure angle	22.44°		22.44°		36°/20°*	
Contact ratio	1.42		1.42		1.19/1.48*	
Gear material	Astaloy Mo +0.25%C, Modulus of Elasticity = 168 GPa, Poisson Ratio = 0.29					
Driving torque, Nm	493	–	493	–	493	–
Contact stress, MPa	1994		2,055 (+3.0%)		1,703 (-14.6%)**	
Root tension stress, MPa	665	630	463(−30.4%)	463(−26.5%)	491(−26.2%)**	461(−26.8%)**
Root compression stress, MPa	847	780	587(−30.7%)	576(−26.2%)	825(−2.6%)**	728(−6.8%)**
Effective contact ratio	1.52		1.42		1.34**	
Transmission error variation, µkm	27.2		15.2 (−44.1)		10.2 (−62.5)**	

Notes
* Drive/coast tooth flanks of the asymmetric gear pair
** The load applied to the drive tooth flank of the asymmetric gear

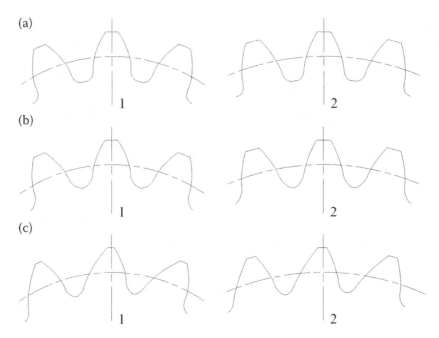

FIGURE 7.3
Driving (1) and driven (2) gear tooth profiles; (a) baseline symmetric, (b) optimized symmetric, (c) optimized asymmetric.

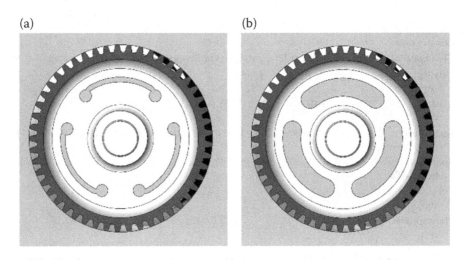

FIGURE 7.4
Hybrid car transmission gear; (a): solid steel machined gear with narrow cut-outs for resonance frequency reduction, (b): PM redesigned gear with reduced resonance frequency, weight, inertia, and fabrication cost. (Courtesy of Höganäs AB, Höganäs, Sweden. With permission.)

a lightweight product, but it also reduces the cost per gear since using less material. In the example in Figure 7.4, the weight reduction is 470 grams, which directly translates to a lower part cost since they are not machined but molded.

For more in-depth information regarding PM materials and processes see [80].

8

Tolerancing and Tolerance Analysis

Tolerances that are seemingly negligibly small in comparison to nominal gear dimensions critically affect gear drive performance and product cost. Incorrect tolerancing can turn a potentially successful project into a total failure.

This chapter considers the tolerance selection approach and shows how tolerancing influences some gear pair performance parameters. It describes the tolerancing and tolerance analysis of asymmetric tooth gears, as a most general case of the directly designed involute gears. Though, it is equally applicable to gears with symmetric teeth.

8.1 Gear Specification

Comprehensive gear drawing specification is essential. With an incompletely specified drawing, a designer practically delegates his responsibilities to a gear supplier, letting him guess about the designer's intentions. The main problem here is that a supplier is not in a position to do this, even if he has gear design experience, because he usually does not know enough about gear drive application specifics. Such guessing may compromise required gear drive performance. The 3D CAD gear model also cannot be a replacement for the appropriately specified gear drawing.

A gear drawing specification table must completely describe gear geometry, including the reference parameters, all critical dimensions with tolerances, an accuracy level defined by the standard accuracy grade or accuracy parameters, and gear inspection dimensions (for example, measurement over/between pins or balls, span measurement, etc.). Standards [81,82] contain examples of proper gear drawing specification tables for most types of involute gears. A gear drawing shows the surface finish, material data including its grade and heat treatment (surface and core hardness, harden case depth), and post-machining surface enhancement (for example, shot peening, superfinishing, coating, etc.). It should also describe gear tooth flank microgeometry shape and parameters, and contain information about tooling, processing, material specimen condition, and some additional information.

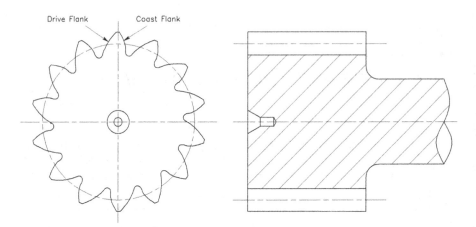

FIGURE 8.1
Asymmetric tooth flank identification in the gear drawing.

Direct Gear Design® specifies the gear dimensions and parameters in a very similar way as for traditionally designed gears to make it understandable to a gear supplier and exclude any guesswork or confusion. Still, such a specification typically may have some additional data. For example, for asymmetric gears, it should have geometric parameters of both drive and coast tooth flanks. Drive and coast tooth flanks of an asymmetric gear should be identified to avoid potential assembly problems (Figure 8.1). A gear drawing should include a gear specification table (see Table 8.1) and a description of the optimized root fillet profile (as a coordinate point table or as part of the CAD tooth profile that accompanies a gear drawing) and its tolerance.

8.2 Accuracy Selection

Gear accuracy selection depends on gear drive application, operating conditions, and technical and market performance requirements. This selection is also affected by chosen gear fabrication technology, materials, heat treatment, etc. The gear handbook [35] describes the types of gear tolerances and their effects on fabrication, cost, and function of a gear drive. For high-performance gear drives, such as, for example, aerospace and racing transmissions, the functional requirements are primary factors for tolerance selection. For less demanding gear drive applications, other factors, such as cost, manufacturability, availability of fabrication equipment, and tooling, can prevail. The gear accuracy standards [83–86] describe tolerance selection.

TABLE 8.1

Gear Specification Table Example

Number of Teeth		a
Module		b
Reference Diameter		b
Pressure Angle at Reference Pitch Diameter	Drive Flank	b
	COAST FLANK	b
Base Diameter	Drive Flank	b
	COAST FLANK	b
Form Diameter	Drive Flank	min/max[a]
	COAST FLANK	min/max[a]
Tooth Tip Diameter		min/max[a]
Root Diameter		min/max[a]
Tooth Thickness at Reference Pitch Diameter		min/max[a]
Tooth Tip Radius		min/max[a]
Face Width		min/max[a]
Pin Diameter		b
Measurement Over Two Pins		min/max[a]
Accuracy Grade Per Agma 2015-A01		b
Runout Tolerance, F_r		a
Total Cumulative Pitch Tolerance, F_p		a
Single Pitch Tolerance, f_{pt}		a
Profile Tolerance	Total, F_α	a
	Form, $f_{f\alpha}$	a
	Slope, $f_{H\alpha}$	a
Helix Tolerance	Total, F_β	a
	Form, $f_{f\beta}$	a
	Slope, $f_{H\beta}$	a
Root Fillet Profile Tolerance		a,c
Mating Gear Part Number		b
Mating Gear Number of Teeth		b
Center Distance		min/max[b]

Notes

a Critical parameter or dimension

b Reference parameter or dimension

c Root fillet profile tolerance is not defined by gear accuracy standards and should be assigned by designer

However, accuracy standards typically do not define tolerances for such dimensions, like, for example, for the tooth tip diameter, root diameter, gear face width, tooth tip radius or chamfer, etc. The standards also do not cover the tooth microgeometry tolerances defining acceptable limits of the tooth profile and lead modifications. They are defined by the gear drive application, operating conditions, experience with the previously designed similar drives, and the prototype performance testing results.

Direct Gear Design® utilizes the same tolerance selection criteria and the same standards as for traditionally designed gears. At the same time, it presents some specific tolerance requirements, for example, the root fillet profile tolerancing. Although the gear tooth root fillet, as an area of maximum bending stress concentration, is critical for gear drive performance, existing gear accuracy standards do not define its tolerances. Its profile and accuracy are marginally defined on the gear drawing by typically a very generous root diameter tolerance and, in some cases, by the minimum fillet radius. Direct Gear Design® optimizes the tooth root fillet profile to minimize the bending stress concentration. It requires a comprehensive specification, tolerancing, and inspection of the tooth root. If the gear tooth profile, including the root fillet, is produced by the same fabrication process and tooling, the fillet profile tolerance can have the same as the involute flank. However, some gear designs require different fabrication methods and separate machining processes and tools for the final shaping of the tooth involute flank and the root fillet. One such gear design assumes using the gear tools (hobs or shaper cutters) with a protuberance for the final machining of the tooth root fillet profile and preliminary machining of the tooth flanks, leaving stock for grinding or shaving after heat treatment (see Section 5.1.2.1). As a result, the involute tooth flank and root fillet accuracy and surface finish are very different. In this case, the accuracy grade achievable by the root fillet profile machining defines its tolerance. Also, the root fillet profile tolerance influences the root diameter tolerance of directly designed gears.

8.3 Tolerance Analysis

The goal of tolerance analysis is to verify the mating gear pair design data in order to guarantee the adequate normal backlash, sufficient root clearance, and adequate contact ratio at any possible tolerance combinations and operating conditions.

Operating conditions include operating temperature and humidity ranges. A wide operating temperature range noticeably changes gear drive dimensions (particularly for large gear transmissions). It also affects the tolerance analysis results in the case of dissimilar materials for gears, gear housing, and other gear drive assembly parts. It is especially critical for gear drives made with polymer components. Besides, some gear polymers like nylons absorb moisture, resulting in increased gear dimensions. Tolerance analysis defines the normal backlash, contact ratio, and root clearances at two extreme (minimum and maximum) value combinations of tolerances and parameters presented in Table 8.2.

TABLE 8.2

Tolerance Analysis Input Data

Dimensions and tolerances	Symbol	Units
Number of teeth of mating gears	z_1, z_2	–
Normal module or diametral pitch at reference pitch diameter	m_n or DP_n	mm or 1/in
Normal pressure angle at reference pitch diameter (for asymmetric gears: drive and coast pressure angles)	α α_d and α_c	°
Helix angle at reference pitch diameter	β	°
Tooth tip diameters (minimum/maximum values)	$d_{amin1}, d_{amax1}, d_{amin2}, d_{amax2}$	mm or in
Root diameters (minimum/maximum values)	$d_{rmin1}, d_{rmax1}, d_{rmin2}, d_{rmax2}$	mm or in
Normal tooth thickness at reference pitch diameter (minimum/maximum values)	$S_{nmin1}, S_{nmax1}, S_{nmin2}, S_{nmax2}$	mm or in
Tooth tip radius (minimum/maximum values)	$R_{amin1}, R_{amax1}, R_{amin2}, R_{amax2}$	mm or in
Gear face width (minimum/maximum values)	$b_{min1}, b_{max1}, b_{min2}, b_{max2}$	mm or in
Runout tolerances[a]	F_{r1}, F_{r2}	mm or in
Single pitch tolerance	f_{pt1}, f_{pt2}	mm or in
Total profile tolerance	$F_{\alpha1}, F_{\alpha2}$	mm or in
Total helix tolerance	$F_{\beta1}, F_{\beta2}$	mm or in
Housing center distance	a_{min}, a_{max}	mm or in
Bearing radial gap	$\delta_{min1}, \delta_{max1}, \delta_{min2}, \delta_{max2}$	mm or in
Operating conditions		
Temperature (minimum/ambient[b]/maximum values)	$T_{min}, T_{amb}, T_{max}$	°C or °F
Humidity (minimum/ambient[b]/maximum values)	$RH_{min}, RH_{amb}, RH_{max}$	%
Material properties		
Linear coefficient of thermal expansion (CTE)	δ_t	mm/mm-°C or in/in-°F
Moisture expansion coefficient (CME)[c]	δ_m	mm/mm-% or in/in-%

Notes

a In some cases the total radial composite deviation (F_i'') that is also known as the total composite error (TTE) is used instead the runout tolerance;

b Ambient temperature or humidity in this content is inspection lab temperature or humidity. Typical ambient temperature is 20°C or 68°F and ambient humidity is about 50%RH;

c For moisture absorbing materials.

(a)   (b)

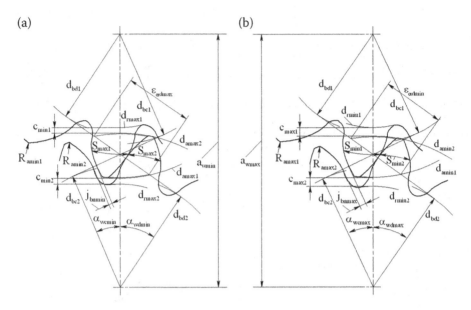

FIGURE 8.2
Case 1 (a) and 2 (b) of extreme tolerance combinations for external gear pair with asymmetric teeth.

Tolerance analysis for an external gear pair considers two cases:

Case 1 (Figure 8.2a): Resulting in minimum normal backlash and radial clearances, and maximum contact ratio. This gear tolerance buildup includes the minimum gear drive housing center distance, bearing radial plays, and tooth tip radii, and the maximum tooth tip diameters, root diameters, and normal tooth thicknesses at reference pitch diameters of mating gears. In this case, runouts of both mating gears should reduce the effective center distance, and the tooth profile, pitch, and helix tolerances increase the effective tooth thickness. Minimum or maximum operating temperature value is selected depending on a gear and housing material combination to reduce the effective center distance. If, for example, the material combination is the steel gears and aluminum housing, the minimum effective center distance is achieved at the minimum operating temperature because the coefficient of thermal expansion (CTE) of steel is lower than that of aluminum. If the material combination is the plastic gears and aluminum housing, the minimum effective center distance is achieved at the maximum operating temperature because the CTE of plastics is usually greater than that of aluminum. Application of moisture-absorbing materials (nylon, for example) for gears and gear drive housing affects the effective center distance. The minimum or maximum value of operating humidity is selected depending on the gear and housing material combination to reduce the effective center distance.

Case 2 (Figure 8.2b): Resulting in the maximum normal backlash and radial clearances, and minimum contact ratio. This gear tolerance buildup includes the maximum housing center distance, bearing radial plays, and tooth tip radii, and the minimum tooth tip diameters, root diameters, and normal tooth thicknesses at reference pitch diameters of mating gears. In this case, runouts of both mating gears should increase the effective center distance, and the tooth profile, pitch, and helix tolerances do not change the effective tooth thickness. Minimum or maximum operating temperature and humidity values are selected depending on gear and housing material combination to increase the effective center distance.

For an internal gear pair, a tolerance analysis approach is the same, but selection of parameter limits is different.

Case 1 (Figure 8.3a): Resulting in the minimum normal backlash and radial clearances, and maximum contact ratio. This gear tolerance buildup includes the maximum housing center distance, minimum bearing radial plays and tooth tip radii, maximum pinion tooth tip and root diameters of the pinion, minimum ring gear tooth tip and root diameters, and maximum normal tooth thickness at reference pitch diameters of mating gears. In this case, the gear runouts should increase the effective center distance, and the tooth profile, pitch, and helix tolerances increase the effective tooth thickness. Minimum or maximum operating temperature and humidity values are selected depending on the gear and housing material combination to increase the effective center distance.

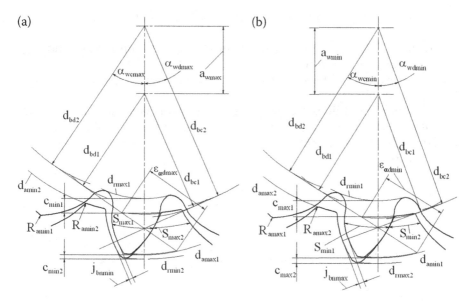

FIGURE 8.3
Case 1 (a) and 2 (b) of extreme tolerance combinations for internal gear pair with asymmetric teeth.

Case 2 (Figure 8.3b): Resulting in the maximum normal backlash and radial clearances, and minimum contact ratio. This gear tolerance buildup includes the minimum housing center distance, maximum bearing radial plays and tooth tip radii, minimum pinion tooth tip and root diameters of the pinion, maximum ring gear tooth tip and root diameters, and minimum normal tooth thickness at reference pitch diameters of mating gears. In this case, the gear runouts should reduce the effective center distance, and the tooth profile, pitch, and helix tolerances combination does not affect the effective tooth thickness. Minimum or maximum operating temperature and humidity values are selected depending on the gear and housing material combination to reduce the effective center distance.

The tolerance analysis set of equations that allows definition of the minimum and maximum values of the normal backlash, radial clearances, and minimum contact ratio is presented below. It is described for helical gears with asymmetric teeth. For spur gears these equations are used assuming the helix angle equals zero. Here is an assumption that the mating gears and gear drive housing are made out of different materials, and those materials could be moisture absorbing. The operating condition coefficients are:

For gears:

$$\lambda_{o1,2} = 1 + \delta_{t1,2}(T_o - T_{amb}) + \delta_{m1,2}(RH_o - RH_{amb}). \tag{8.1}$$

For gear drive housing:

$$\lambda_{oh} = 1 + \delta_{th}(T_o - T_{amb}) + \delta_{mh}(RH_o - RH_{amb}), \tag{8.2}$$

where T_{amb} and RH_{amb} are ambient temperature and humidity, T_o and RH_o are operating temperature and humidity, which can have minimum, ambient, or maximum values. Symbols 1, 2, and h are for the pinion, gear, and housing, accordingly.

For ambient conditions coefficients $\lambda_{o1,2}$ and λ_{oh} are equal to 1.0. For actual operating conditions major gear and housing dimensions should be adjusted by these coefficients:

Center distance:

$$a_{\min} = \lambda_{oh} \times a_{\min} \text{ and } a_{\max} = \lambda_{oh} \times a_{\max}. \tag{8.3}$$

Tooth tip diameters:

$$d_{a\,\min\,1,2} = \lambda_{o1,2} \times d_{a\,\min\,1,2} \text{ and } d_{a\,\max\,1,2} = \lambda_{o1,2} \times d_{a\,\max\,1,2}. \tag{8.4}$$

Root diameters:

$$d_{r\,\min\,1,2} = \lambda_{o1,2} \times d_{r\,\min\,1,2} \text{ and } d_{r\,\max\,1,2} = \lambda_{o1,2} \times d_{r\,\max\,1,2}. \tag{8.5}$$

Normal tooth thicknesses:

$$s_{n \min 1,2} = \lambda_{o1,2} \times s_{n \min 1,2}$$

and

$$s_{n \max 1,2} = \lambda_{o1,2} \times s_{n \max 1,2} + f_{pt1,2} + F_{\alpha1,2} + F_{\beta1,2}. \tag{8.6}$$

Gear face widths:

$$b_{\min 1,2} = \lambda_{o1,2} \times b_{\min 1,2} \quad \text{and} \quad b_{\max 1,2} = \lambda_{o1,2} \times b_{\max 1,2}. \tag{8.7}$$

It is not necessary to apply operating condition coefficients $\lambda_{o1,2}$ and λ_{oh} to small dimensions like the tooth tip radii or gear and gear housing accuracy tolerances, because typically it does not make any noticeable difference.

Then the operating center distance values $a_{w\min}$ and $a_{w\max}$ are:
For external gearing:

$$a_{w \min} = a_{\min} - \frac{F_{r1}}{2} - \frac{F_{r2}}{2} + \frac{\delta_{\min 1}}{2} + \frac{\delta_{\min 2}}{2}$$

and

$$a_{w \max} = a_{\max} + \frac{F_{r1}}{2} + \frac{F_{r2}}{2} + \frac{\delta_{\max 1}}{2} + \frac{\delta_{\max 2}}{2}. \tag{8.8}$$

For internal gearing:

$$a_{w \min} = a_{\min} - \frac{F_{r1}}{2} - \frac{F_{r2}}{2} - \frac{\delta_{\max 1}}{2} - \frac{\delta_{\max 2}}{2}$$

and

$$a_{w \max} = a_{\min} + \frac{F_{r1}}{2} + \frac{F_{r2}}{2} - \frac{\delta_{\min 1}}{2} - \frac{\delta_{\min 2}}{2}. \tag{8.9}$$

The reference pitch diameters $d_{p1,2}$ are:
In the Metric system:

$$d_{p1,2} = \frac{m_n z_{1,2} \lambda_{o1,2}}{\cos \beta}. \tag{8.10}$$

In the English system:

$$d_{p1,2} = \frac{z_{1,2}\lambda_{o1,2}}{DP_n \cos\beta}.$$ (8.11)

The transverse pressure angles α_{td} and α_{tc} at the reference pitch diameters are:
For drive flanks:

$$\alpha_{td} = \arctan\left(\frac{\tan\alpha_d}{\cos\beta}\right).$$ (8.12)

For coast flanks:

$$\alpha_{tc} = \arctan\left(\frac{\tan\alpha_c}{\cos\beta}\right).$$ (8.13)

The base diameters $d_{b1,2}$ are:
For drive flanks:

$$d_{bd1,2} = d_{p1,2}\cos\alpha_{td}.$$ (8.14)

For coast flanks:

$$d_{bc1,2} = d_{p1,2}\cos\alpha_{tc}.$$ (8.15)

The helix angles β_{bd} and β_{bc} at the base diameters are:
For drive flanks:

$$\beta_{bd} = \arctan(\tan\beta\cos\alpha_{td}).$$ (8.16)

For coast flanks:

$$\beta_{bc} = \arctan(\tan\beta\cos\alpha_{tc}).$$ (8.17)

The tooth tip involute angles $\alpha_{admin1,2}$ and $\alpha_{admax1,2}$ are

$$\alpha_{ad\,min\,1,2} = \arccos\left(\frac{d_{bd1,2}}{d_{a\,min\,1,2}}\right)$$

and

$$\alpha_{ad\,max\,1,2} = \arccos\left(\frac{d_{bd1,2}}{d_{a\,max\,1,2}}\right).$$ (8.18)

The helix angles at the tooth tip diameters $\beta_{amin1,2}$ and $\beta_{amax1,2}$ are

$$\beta_{a\ min\ 1,2} = \arctan\left(\tan\beta\frac{\cos\alpha_{td}}{\cos\alpha_{ad\ min\ 1,2}}\right)$$

and

$$\beta_{a\ max\ 1,2} = \arctan\left(\tan\beta\frac{\cos\alpha_{td}}{\cos\alpha_{ad\ max\ 1,2}}\right). \tag{8.19}$$

The effective tooth tip involute angles for gears with external teeth from (2.40) and (2.41) are:

For drive flanks $\alpha_{edmin1,2}$ and $\alpha_{edmax1,2}$:

$$\alpha_{ed\ min\ 1,2} = \arctan\left(\tan\left(\arccos\left(\frac{d_{bd1,2}}{d_{a\ min\ 1,2} - 2R_{a\ max\ 1,2}}\right)\right) + \frac{2R_{a\ max\ 1,2}}{d_{bd1,2}}\right)$$

and

$$\alpha_{ed\ max\ 1,2} = \arctan\left(\tan\left(\arccos\left(\frac{d_{bd1,2}}{d_{a\ max\ 1,2} - 2R_{a\ min\ 1,2}}\right)\right) + \frac{2R_{a\ min\ 1,2}}{d_{bd1,2}}\right). \tag{8.20}$$

For coast flanks $\alpha_{ecmin1,2}$ and $\alpha_{ecmax1,2}$:

$$\alpha_{ec\ min\ 1,2} = \arctan\left(\tan\left(\arccos\left(\frac{d_{bc1,2}}{d_{a\ min\ 1,2} - 2R_{a\ max\ 1,2}}\right)\right) + \frac{2R_{a\ max\ 1,2}}{d_{bc1,2}}\right)$$

and

$$\alpha_{ec\ max\ 1,2} = \arctan\left(\tan\left(\arccos\left(\frac{d_{bc1,2}}{d_{a\ max\ 1,2} - 2R_{a\ min\ 1,2}}\right)\right) + \frac{2R_{a\ min\ 1,2}}{d_{bc1,2}}\right). \tag{8.21}$$

The tooth tip effective involute angles for a gear with internal teeth from (2.40) and (2.41) are:

For drive flanks $\alpha_{edmin1,2}$ and $\alpha_{edmax1,2}$:

$$\alpha_{ed\ min\ 2} = \arctan\left(\tan\left(\arccos\left(\frac{d_{bd2}}{d_{a\ min\ 2} + 2R_{a\ max\ 2}}\right)\right) - \frac{2R_{a\ max\ 2}}{d_{bd2}}\right)$$

and

$$\alpha_{aed\ max\ 2} = \arctan\left(\tan\left(\arccos\left(\frac{d_{bd2}}{d_{a\ max\ 2} + 2R_{a\ min\ 2}}\right)\right) - \frac{2R_{a\ min\ 2}}{d_{bd2}}\right). \quad (8.22)$$

For coast flanks $\alpha_{ecmin1,2}$ and $\alpha_{ecmax1,2}$:

$$\alpha_{ec\ min\ 2} = \arctan\left(\tan\left(\arccos\left(\frac{d_{bc2}}{d_{a\ min\ 2} + 2R_{a\ max\ 2}}\right)\right) - \frac{2R_{a\ max\ 2}}{d_{bc2}}\right)$$

and

$$\alpha_{ec\ max\ 2} = \arctan\left(\tan\left(\arccos\left(\frac{d_{bc2}}{d_{a\ max\ 2} + 2R_{a\ min\ 2}}\right)\right) - \frac{2R_{a\ min\ 2}}{d_{bc2}}\right). \quad (8.23)$$

From (2.24) the asymmetry factor K is

$$K = d_{bc1,2}/d_{bd1,2} = \cos\alpha_{tc}/\cos\alpha_{td}. \quad (8.24)$$

The involute intersection profile angles $\nu_{dmin1,2}$ and $\nu_{dmax1,2}$ for gears with external teeth are defined from a system of equations:

$$inv(\nu_{d\ min\ 1,2}) + inv(\nu_{c\ min\ 1,2}) = inv(\alpha_{td}) + inv(\alpha_{tc}) + \frac{2s_{n\ min\ 1,2}\tan\alpha_{td}}{d_{bd1,2}\cos\beta}. \quad (8.25)$$

and

$$\cos\nu_{c\ min\ 1,2} = K\cos\nu_{d\ min\ 1,2}, \quad (8.26)$$

$$inv(\nu_{d\ max\ 1,2}) + inv(\nu_{c\ max\ 1,2}) = inv(\alpha_{td}) + inv(\alpha_{tc}) + \frac{2s_{n\ max\ 1,2}\tan\alpha_{td}}{d_{bd1,2}\cos\beta} \quad (8.27)$$

and

$$\cos\nu_{c\ max\ 1,2} = K\cos\nu_{d\ max\ 1,2}. \quad (8.28)$$

The involute intersection profile angles ν_{dmin2} and ν_{dmax2} for gears with internal teeth are defined from a system of equations:

$$inv\,(v_{d\,min\,2}) + inv\,(v_{c\,min\,2}) = inv\,(\alpha_{td}) + inv\,(\alpha_{tc}) + \frac{2(P_n - s_{n\,max\,1,2})\tan\alpha_{td}}{d_{bd2}\cos\beta}$$

(8.29)

and (8.24), and

$$inv\,(v_{d\,max\,2}) + inv\,(v_{c\,max\,2}) = inv\,(\alpha_{td}) + inv\,(\alpha_{tc}) + \frac{2(P_n - s_{n\,min\,2})\tan\alpha_{td}}{d_{bd2}\cos\beta}$$

(8.30)

and (8.26), where P_n is a normal circular pitch at the reference pitch diameter that is equal to $P_n = \pi m_n$ or $P_n = \pi/DP_n$ in the Metric or English system, accordingly.

Then operating pressure angles are:

For drive flanks $\alpha_{wd\,min}$ and $\alpha_{wd\,max}$:

$$\alpha_{wd\,min} = \arccos\left(\frac{d_{bd1}(u \pm 1)}{2a_{wmin}}\right)$$

and

$$\alpha_{wd\,max} = \arccos\left(\frac{d_{bd1}(u \pm 1)}{2a_{wmax}}\right),$$

(8.31)

where $u = z_2/z_1$ is gear ratio, and $+$ is for the external gearing and $-$ is for the internal gearing.

For coast flanks $\alpha_{wc\,min}$ and $\alpha_{wc\,max}$:

$$\alpha_{wc\,min} = \arccos(K \cos \alpha_{wd\,min})$$

and

$$\alpha_{wc\,max} = \arccos(K \cos \alpha_{wd\,max}).$$

(8.32)

The operating transverse tooth thicknesses $S_{wmin1,2}$ and $S_{wmax1,2}$ are:

For gears with external teeth:

$$S_{w\,min\,1,2} = \frac{d_{bd1,2}}{2\cos\alpha_{wd\,max}}(inv\,(v_{d\,min\,1,2}) + inv\,(v_{c\,min\,1,2}) - inv\,(\alpha_{wd\,max})$$
$$- inv\,(\alpha_{wc\,max}))$$

and

$$s_{w \, max \, 1,2} = \frac{d_{bd1,2}}{2 \cos \alpha_{wd \, min}} (inv(\nu_{d \, max \, 1,2}) + inv(\nu_{c \, max \, 1,2}) - inv(\alpha_{wd \, min}) \quad (8.33)$$
$$- inv(\alpha_{wc \, min})).$$

For gears with internal teeth:

$$s_{w \, min \, 2} = \frac{d_{bd2}}{2 \cos \alpha_{wd \, min}} \left(\frac{2\pi}{z_2} - inv(\nu_{d \, max \, 2}) - inv(\nu_{c \, max \, 2}) + inv(\alpha_{wd \, min}) \right.$$
$$\left. + inv(\alpha_{wc \, min}) \right)$$

and

$$s_{w \, max \, 2} = \frac{d_{bd2}}{2 \cos \alpha_{wd \, max}} \left(\frac{2\pi}{z_2} - inv(\nu_{d \, min \, 2}) - inv(\nu_{c \, min \, 2}) + inv(\alpha_{wd \, max}) \right. \quad (8.34)$$
$$\left. + inv(\alpha_{wc \, max}) \right).$$

For gears with asymmetric teeth the normal backlash is defined between the coast tooth flanks, when the drive tooth flanks are in contact. The normal backlash values $j_{bnmin1,2}$ and $j_{bnmax1,2}$ are

$$j_{bn \, min} = \left(\frac{\pi d_{bd1,2}}{z_1 \cos \alpha_{wd \, min}} - s_{w \, max \, 1} - s_{w \, max \, 2} \right) \cos \alpha_{wc \, min} \cos \beta_{bc}$$

and

$$j_{bn \, max} = \left(\frac{\pi d_{bd1,2}}{z_1 \cos \alpha_{wd \, max}} - s_{w \, min \, 1} - s_{w \, min \, 2} \right) \cos \alpha_{wc \, max} \cos \beta_{bc}. \quad (8.35)$$

The normal backlash must be greater than the maximum possible gear tooth tip deflection under the load. This allows avoidance of simultaneous contact of the opposite tooth flanks with mating gear tooth flanks.

The root clearances $c_{min1,2}$ and $c_{max1,2}$ are:

For external gearing:

$$c_{min \, 1} = a_{min} - \frac{d_{a \, min \, 2}}{2} - \frac{d_{r \, max \, 1}}{2} \quad \text{and} \quad c_{max \, 1} = a_{max} - \frac{d_{a \, max \, 2}}{2} - \frac{d_{r \, min \, 1}}{2},$$
$$(8.36)$$

$$c_{min \, 2} = a_{min} - \frac{d_{a \, min \, 1}}{2} - \frac{d_{r \, max \, 2}}{2} \quad \text{and} \quad c_{max \, 2} = a_{max} - \frac{d_{a \, max \, 1}}{2} - \frac{d_{r \, min \, 2}}{2}.$$
$$(8.37)$$

For internal gearing:

$$c_{\min 1} = \frac{d_{a\,\min 2}}{2} - \frac{d_{r\,\max 1}}{2} - a_{\max} \text{ and } c_{\max 1} = \frac{d_{a\,\max 2}}{2} - \frac{d_{r\,\min 1}}{2} - a_{\min},$$

$$(8.38)$$

$$c_{\min 2} = \frac{d_{r\,\min 2}}{2} - \frac{d_{a\,\max 1}}{2} - a_{\max} \text{ and } c_{\max 2} = \frac{d_{r\,\max 2}}{2} - \frac{d_{a\,\min 1}}{2} - a_{\min}.$$

$$(8.39)$$

The minimal root clearances must be greater than zero to avoid tooth tip/root interference. The tooth root fillet construction and optimization (see Section 5.1.2) excludes the tooth tip/root interference. However, operating conditions, including temperature and humidity, may reduce root clearances. Besides, low radial clearances may result in trapping lubricant in the tooth root area, increased hydraulic losses, and reduced gear efficiency, especially for relatively wide spur gears. This may require designing the tooth fillet with increased root clearances even with some compromise of bending stress reduction.

The transverse operating contact ratios are:

For drive flanks of external gear mesh $\varepsilon_{ad\min}$ and $\varepsilon_{ad\max}$:

$$\varepsilon_{ad\,\min} = \frac{z_1}{2\pi}(\tan\alpha_{aed\,\min 1} + u\tan\alpha_{aed\,\min 2} - (u+1)\tan\alpha_{wd\,\max})$$

and

$$\varepsilon_{ad\,\max} = \frac{z_1}{2\pi}(\tan\alpha_{aed\,\max 1} + u\tan\alpha_{aed\,\max 2} - (u+1)\tan\alpha_{wd\,\min}). \quad (8.40)$$

For coast flanks of external gear mesh $\varepsilon_{ac\min}$ and $\varepsilon_{ac\max}$:

$$\varepsilon_{ac\,\min} = \frac{z_1}{2\pi}(\tan\alpha_{aec\,\min 1} + u\tan\alpha_{aec\,\min 2} - (u+1)\tan\alpha_{wc\,\max})$$

and

$$\varepsilon_{ac\,\max} = \frac{z_1}{2\pi}(\tan\alpha_{aec\,\max 1} + u\tan\alpha_{aec\,\max 2} - (u+1)\tan\alpha_{wc\,\min}). \quad (8.41)$$

For drive flanks of internal gear mesh $\varepsilon_{ad\min}$ and $\varepsilon_{ad\max}$:

$$\varepsilon_{ad\,\min} = \frac{z_1}{2\pi}(\tan\alpha_{aed\,\max 1} - u\tan\alpha_{aed\,\min 2} + (u-1)\tan\alpha_{wd\,\min})$$

and

$$\varepsilon_{ad\ max} = \frac{z_1}{2\pi}(\tan\alpha_{aed\ min\ 1} - u\tan\alpha_{aed\ max\ 2} + (u-1)\tan\alpha_{wd\ max}). \quad (8.42)$$

For coast flanks of internal gear mesh ε_{acmin} and ε_{acmax}:

$$\varepsilon_{ac\ min} = \frac{z_1}{2\pi}(\tan\alpha_{aec\ max\ 1} - u\tan\alpha_{aec\ min\ 2} + (u-1)\tan\alpha_{wc\ min})$$

and

$$\varepsilon_{ac\ max} = \frac{z_1}{2\pi}(\tan\alpha_{aec\ min\ 1} - u\tan\alpha_{aec\ max\ 2} + (u-1)\tan\alpha_{wc\ max}). \quad (8.43)$$

The axial operating contact ratio values $\varepsilon_{\beta min}$ and $\varepsilon_{\beta max}$ from (2.100) are

$$\varepsilon_{\beta\ min} = \frac{b_{w\ min}\tan\beta_{bd}}{p_{bd}} = \frac{b_{w\ min}\tan\beta_{bc}}{p_{bc}}$$

and

$$\varepsilon_{\beta\ max} = \frac{b_{w\ max}\tan\beta_{bd}}{p_{bd}} = \frac{b_{w\ min}\tan\beta_{bc}}{p_{bc}}, \quad (8.44)$$

where p_{bd} and p_{bc} are transverse base circular pitches of the drives and coast tooth flanks, accordingly, and b_{wmin} and b_{wmax} are the minimum and maximum values of the operating face width of gear engagement (see Figure 8.4).

In asymmetric gearing the operating axial contact ratio is the same for the drive and coast tooth flanks.

The total operating contact ratios are defined from Equations (2.93) and (2.94):

For drive flanks $\varepsilon_{\gamma dmin}$ and $\varepsilon_{\gamma dmax}$:

$$\varepsilon_{\gamma d\ min} = \varepsilon_{ad\ min} + \varepsilon_{\beta\ min}\ \text{and}\ \varepsilon_{\gamma d\ max} = \varepsilon_{ad\ max} + \varepsilon_{\beta\ max}. \quad (8.45)$$

For coast flanks $\varepsilon_{\gamma cmin}$ and $\varepsilon_{\gamma cmax}$:

$$\varepsilon_{\gamma c\ min} = \varepsilon_{ac\ min} + \varepsilon_{\beta\ min}\ \text{and}\ \varepsilon_{\gamma c\ max} = \varepsilon_{ac\ max} + \varepsilon_{\beta\ max}. \quad (8.46)$$

The minimum total operating contact ratio must be greater than 1.0 for smooth tooth pair engagement.

The equations above describe the absolute tolerance analysis approach with output parameters (normal backlash, radial clearances, and contact ratio) defined at two extreme tolerance combinations, described above as

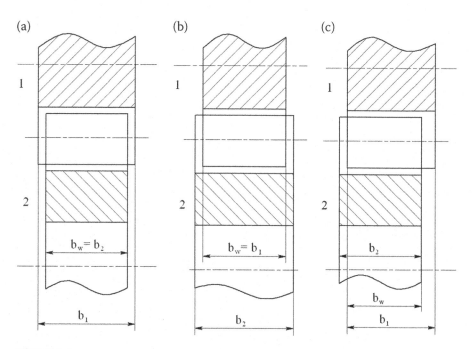

FIGURE 8.4
Operating face width b_w definition; (a): $b_w = b_2$, (b): $b_w = b_1$, (c): gears are assembled with axial shift.

Cases 1 and 2. As a result, differences between these parameters' minimum and maximum values can be considerable. At the same time, a substantial number of dimensions involved in a tolerance analysis create a very low probability of simultaneous coincidence of extreme tolerance combinations of their minimum or maximum values. Application of the statistical tolerance analysis allows the reduction of tolerance ranges of output parameters or achievement of required ranges with the larger input tolerances, which reduces production cost. According to paper [87], "statistical tolerance analysis can be used by designers and manufacturing personnel to take advantage of statistical averaging over assemblies of parts, allowing the use of less restrictive tolerances in exchange for admitting the small probability of non-assembly." The reduction of this probability can be provided by the statistical process control (SPC) that establishes the technological tolerances for critical dimensions that are lower than the drawing tolerances. The publications [88,89] describe the statistical tolerance analysis approach.

9

Gear Manufacturing Essentials

9.1 Fabrication Technologies and Tooling

Direct Gear Design® provides improved performance for custom gear drives, achieved by gear tooth macro- and microgeometry enhancement, which also requires some customization of manufacturing technology and tooling. This chapter considers some important fabrication issues of directly designed gears.

There are many ways to fabricate gears. They can be divided into three main groups:

- Machining processes that shape a gear by material removal
- Forming processes that shape a gear by mechanical deformation of material or changing material state
- Additive processes that shape a gear by material deposit

Other manufacturing processes used for gear fabrication, like heat treatment, surface engineering, coating, etc., are not considered in this book.

This chapter describes the manufacturing specifics of asymmetric tooth gears as the most general form of directly designed involute gears. However, it is equally applicable to gears with symmetric teeth.

9.1.1 Gear Machining

Table 9.1 presents main machining processes used for spur and helical gears.

9.1.1.1 Form Machining

A form gear cutting or grinding tool profile (Figure 9.1) is the same as a space profile between gear teeth. This process is applicable for spur and helical [90] gears. A form machining tool is unique for every gear tooth profile, and its cost is practically the same for standard or custom directly

TABLE 9.1

Spur and Helical Gear Machining Processes

Type of process		Type of tooling
Form machining	Cutting	Form disk or end mill cutter, broach, etc.
	Grinding	Grinding wheel
Generating machining	Cutting	Hob, shaper cutter, rack cutter, shaver cutter
	Grinding	Grinding wheel
Contour machining	CNC milling	Cylinder or ball mill cutter
	Wire cut EDM*	Wire
	Laser cutting	Laser beam
	Water Jet cutting	Water or water and abrasive media mixture stream

Note
* EDM – electric discharge machining

designed gears with similar geometry. Accuracy of form cutting tool positioning relative to the gear blank is critically important.

The end mill cutter can be used for form machining only symmetric tooth gears (Figure 9.2). Gears with a low number of teeth and the optimized root fillet profile may have an undercut below the form circle. In this case, two form disk cutters (Figure 9.3) can be used to machine such tooth profiles. The disk cutter profiles overlap the root fillet at its bottom, which may result in a little step.

The gear form machining process typically removes material from one space between teeth after another. When the machining of one space between teeth is completed, the indexing device (usually a rotary table) positions a gear blank for the next tooth space cutting. Figure 9.4 presents asymmetric the gear form grinding setup with cubic boron nitride (CBN) form grinding wheels.

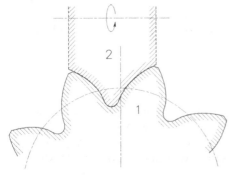

FIGURE 9.1
Gear form machining; 1: gear profile (solid contour); 2: tool profile (dashed contour).

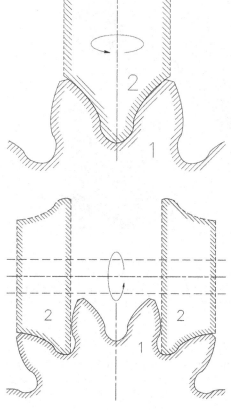

FIGURE 9.2
End mill cutter machining; 1: gear profile (solid contour); 2: cutter profile (dashed contour).

FIGURE 9.3
Machining of teeth with undercut root fillet; 1: gear profile (solid contour); 2: form disk cutters (dashed contour).

One type of form machining processes – gear fly cutting [91] – has a tool (a fly cutter) in a mesh with the gear blank (Figures 9.5 and 9.6). Fly cutting uses conventional gear hobbing machines that operate when a hob cutter and a gear blank are in constant synchronized rotation. However, unlike the gear rack generating process, a space between teeth is shaped by a cutter tooth profile as in conventional form machining with the disk cutter (Figure 9.5a).

The gear fly cutter also looks similar to the conventional gear disk cutter, but all cutter edges are turned at the start angle φ, which is defined as follows:

$$\phi = \arcsin\frac{m_n n_t}{d_t}, \tag{9.1}$$

where m_n is the normal module of the gear, n_t is the number of fly cutter teeth, and d_t is the fly cutter pitch diameter.

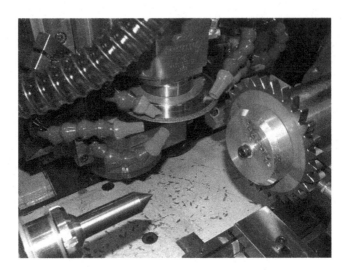

FIGURE 9.4
Asymmetric gear form grinding setup with CBN grinding wheel. (Courtesy of Aero Gear, South Windsor, Connecticut. From Brown, F.W., et al. Analysis and testing of gears with asymmetric involute tooth form and optimized fillet form for potential application in helicopter main drives, *AGMA Fall Technical Meeting*, Milwaukee, Wisconsin, October 18–19, 2010, 10FTM14. With permission.)

A gear fly cutter can also be considered a multi-start gear hob that has just one tooth in each start. The cutter setup angle ϕ relative to the gear plane is defined the same way as for conventional gear hobbing:

$$\varphi = \beta \pm \phi, \tag{9.2}$$

where β is the gear helix angle, the + sign indicates the gear helix and cutter start angle have the same directions (right-right or left-left), and a – sign indicates the gear helix and cutter start angle have opposite directions (right-left or right-left).

Figure 9.6 shows the adjustable gear fly cutter. This makes it possible to machine symmetric and asymmetric gears with different numbers of teeth and modules by using replaceable cutting and angle inserts. Such a gear fly cutter is very convenient for prototyping practically any custom tooth shape gears.

9.1.1.2 Generating Machining

Figure 9.7 shows the schematics of the rack generating asymmetric gear machining process and asymmetric gear hob cutter. In this process, a gear blank and tool are engaged in a mesh, and all gear teeth are machined simultaneously. In traditional gear design, the tooling (a hob or rack cutter)

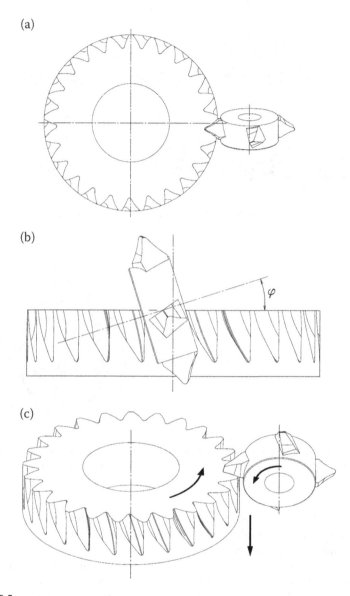

(a)

(b)

φ

(c)

FIGURE 9.5
Gear fly cutting schematics: (a) top view, (b) right view, (c) isometric view; φ: cutter setup angle.

profile is known at the beginning of gear design. In combination with its position relative to the gear center, established by the addendum modification (or X-shift), the tooling rack profile defines the gear tooth profile.

Direct Gear Design® describes the gear tooth geometry without using any preselected tooling rack and its position relatively a gear center. On the contrary, a tooling rack contour is defined by an already known completely

(a)   (b)

FIGURE 9.6
Gear fly cutting: (a) hobbing machine setup, (b) adjustable fly cutter with the cutting inserts.

(a)   (b)

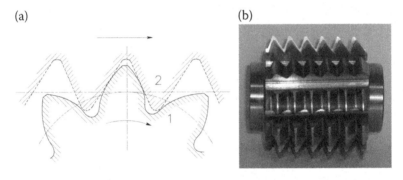

FIGURE 9.7
Schematics of the gear rack generating and gear hob cutter; (a) rack generating, 1: gear profile (solid contour), 2: hob or rack cutter profile (dashed contour); (b) asymmetric gear hob cutter.

optimized gear tooth profile applying the reverse generating technique. This technique assumes that every point of the gear tooth profile should have its mating point on the rack tooth contour.

Figure 9.8 demonstrates how the tooling rack profile point A_t position is defined from the gear tooth profile point A_g position. To find the generating rack profile point A_t corresponding to the gear tooth profile point A_g, the line A_gB_g perpendicular to the tooth profile in point A_g is constructed. The point B_g lies at an intersection of the line A_gB_g with the gear pitch circle. The gear tooth profile and line A_gB_g are rotated on the angle γ_g relative to the gear center until point B_g reaches its pitch point position B_g', where the gear pitch circle 3 is tangent to the rack pitch line 4. Then the point A_g is in the position A_g', where the gear tooth profile 1' is tangent to the rack profile 2'. The line $A_g'B_g'$ is moved parallel to the rack pitch line 4 on the distance $B_g'B_t$, which is equal to the length of the arc B_gB_g'. This movement puts the point A_g'

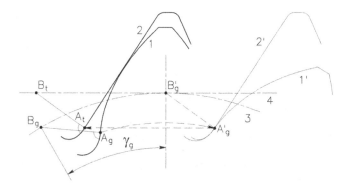

FIGURE 9.8
Generating rack profile definition; 1 and 1′: gear profile positions; 2 and 2′: rack cutter profile positions; 3: gear pitch circle in mesh with rack; 4: rack pitch line in mesh with gear.

in the position A_t at the rack profile that corresponds to the point A_g at the gear tooth profile. This approach allows the definition of any generating rack profile point that is related to a certain gear tooth profile point.

Figure 9.9 shows the asymmetric pump gears and the hob, whose rack profile is defined by reverse generation.

Some gear drives, including automotive transmissions, utilize gears that are machined with protuberance hobs that form the final tooth root shape leaving a required grinding stock at the involute flank. After the hobbing operation, these gears have an undercut between the root fillet preliminary machined tooth flank. Subsequently, this undercut is partially removed by the tooth flank grinding (or shaving) after the heat treatment (typically carburizing) and, in some cases, a root shot peening that results in compressive residual stress that strengthens the tooth root.

Figure 9.10 shows the reverse generation of the topping protuberance hob rack profile. It is similar to the reverse generation of the conventional gear

(a)   (b)

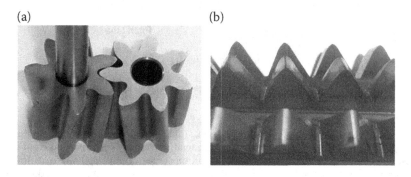

FIGURE 9.9
A sample of the gear rack reverse generation; (a) asymmetric pump gears, (b) hob teeth (Courtesy of Melling Engine Parts, Jackson, Michigan, USA. With permission.)

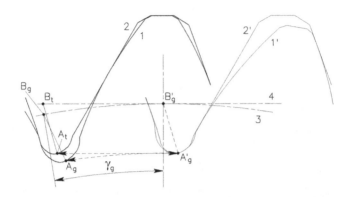

FIGURE 9.10
Reverse generating of the protuberance hob rack profile; 1 and 1′: gear profile positions; 2 and 2′: rack cutter profile positions; 3: gear pitch circle in a mesh with the rack; 4: rack pitch line in a mesh with the gear.

hob rack (Figure 9.8), except for the protuberance tip to form the gear tooth root undercut and the chamfer edges for the gear tooth tip chamfers.

The gear pitch diameter in the mesh with a generating rack is not necessarily equal to the operating pitch diameter with the mating gear. Its selection is very important, because it affects all protuberance hob profile parameters (Figure 9.11). If gear pitch diameter in the mesh with a generating rack is greater than the nominal gear pitch diameter in the mesh

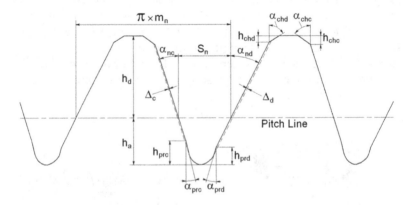

FIGURE 9.11
Topping protuberance hob profile parameters; m_n: normal module, S_n: normal tooth thickness at the generating pitch line, h_a: addendum, h_d: dedendum, α_{nd} and α_{nc}: drive and coast normal pressure angles, h_{prd} and h_{prc}: drive and coast protuberance heights, α_{prd} and α_{prc}: drive and coast protuberance angles, h_{chd} and h_{chc}: drive and coast chamfer heights, α_{chd} and α_{chc}: drive and coast chamfer angles, Δ_d and Δ_c: drive and coast tooth flank grinding stocks. (Kapelevich A.L. and Y.V. Shekhtman. Optimization of asymmetric gear tooth root generated with protuberance hob, *Gear Technology*. June 2020, 32–37. With permission.)

with the mating gear, the hob rack normal module m_n, α_{nd} and α_{nc} (drive and coast pressure angles) and α_{prd} and α_{prc} (drive and coast protuberance angles) are increased. This leads to a hob protuberance size reduction that may result in increased root fillet surface roughness and excessive hob protuberance tip wear. If gear pitch diameter in the mesh with a generating rack is smaller than the nominal gear pitch diameter in the mesh with the mating gear, the hob rack module m_n, α_{nd} and α_{nc} (drive and coast pressure angles) and α_{prd} and α_{prc} (drive and coast protuberance angles) are reduced. This leads to a hob protuberance size increase and a smoother root fillet surface finish. However, the drive and coast protuberance angles should not be less than about 8°, otherwise the back angle of the protuberance cutting edge could be too small for effective cutting.

Figure 9.12 shows the asymmetric protuberance hob cutter, tooth flank grinding wheel, and asymmetric gear.

In some cases, a combination of geometric parameters does not allow the use of a protuberance hob for preliminary machining of the tooth flanks and the final machining of the root fillet. It is typical for gears with a low number of teeth when a sweep of the protuberance hob tooth tip undercuts the tooth flanks near the form diameters (Figure 9.13).

In this case, the gear machining process should be altered, replacing protuberance hobbing with conventional non-protuberance hobbing (Figure 9.14a) and root milling using a mill cutter shaped as the optimized root fillet (Figure 9.14b). After heat treatment, the gear flanks are ground

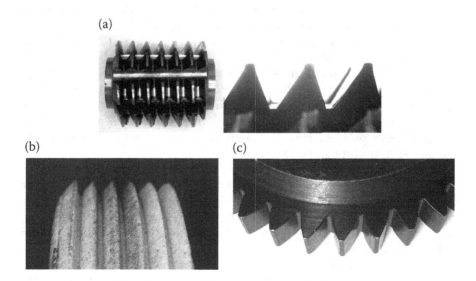

(a)

(b) (c)

FIGURE 9.12
(a) Asymmetric protuberance hob cutter and its teeth (magnified), (b) generating helical grinding wheel, (c) asymmetric tooth gear.

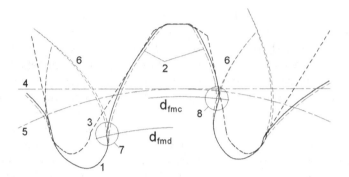

FIGURE 9.13
Gear profile undercut by protuberance hob tooth tip; 1: machined gear profile with the grinding stock, 2: final gear flanks profiles after grinding, 3: protuberance hob profile; 4: hobbing generating pitch line; 5: hobbing generating pitch diameter; 6: hob tooth tip trajectories; 7 and 8: drive and coast flank undercut areas; d_{fmd} and d_{fmc}: drive and coast flank form diameters of the gear tooth with the grinding stock. (Kapelevich A.L. and Y.V. Shekhtman. Optimization of asymmetric gear tooth root generated with protuberance hob, *Gear Technology*. June 2020, 32–37. With permission.)

with a helical grinding wheel. Asymmetric gears have different drive and coast flank form diameters, and the helical grinding wheel thread tip should be dressed with a chamfer angle f to avoid interference with the root fillet near the coast flank form diameter (Figure 9.14c). This gear machining process, combining the generating non-protuberance hobbing, form milling of the optimized root fillet, and final generating grinding of the involute flanks, was used to manufacture the asymmetric pinion of the helicopter gearbox described in Section 11.3.

If a gear is to be made by shaper cutting, the traditional gear design suggests that the shaper cutter parameters and profile should be known before gear design. In combination with its position relative to the gear center (addendum modification or X-shift), the shaper cutter profile defines the gear tooth profiles. Figure 9.15 shows the schematics of the shaper generating gear cutting process.

Direct Gear Design® defines a shaper cutter profile also after the gear profile is already known. The gear profile is used for the shaper cutter profile reverse generation utilizing a technique similar to that for the rack tooling profile definition. It assumes that in the shaper cutter/gear mesh, every point of the gear tooth profile has its mating point on the shaper cutter tooth profile.

Figure 9.16 demonstrates how the shaper cutter profile point A_t position is defined from the gear tooth profile point A_g position. To find the shaper cutter profile point A_t corresponding to the gear tooth profile point Ag, the line A_gB_g perpendicular to the tooth profile is constructed. Point B_g lies in an

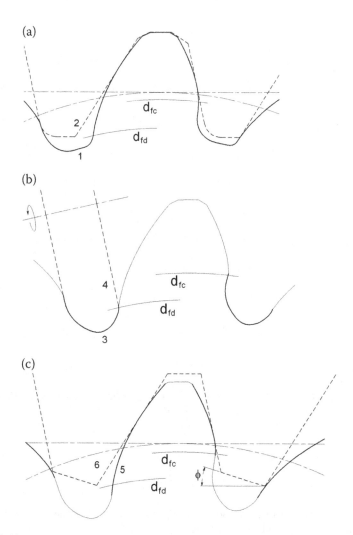

FIGURE 9.14
Processing of asymmetric gear with low number of teeth; (a) preliminary gear hobbing, 1: preliminary machined gear profile, 2: roughing non-protuberance hob profile; (b) optimized root milling, 3: gear with machined root fillet, 4: form disk mill cutter; (c) tooth flank grinding, 5: gear with ground tooth flanks, 6: generating helical grinding wheel; ϕ: grinding wheel thread tip angle. (Kapelevich A.L. and Y.V. Shekhtman. Optimization of asymmetric gear tooth root generated with protuberance hob, *Gear Technology*. June 2020, 32–37. With permission.)

intersection of the line $A_g B_g$ with the gear pitch circle. The gear tooth profile and line $A_g B_g$ are rotated on the angle γ_g relative to the gear center until point B_g reaches its pitch point position B_g', where gear pitch circle 3 is tangent to shaper cutter pitch circle 4. The angle γ_g is

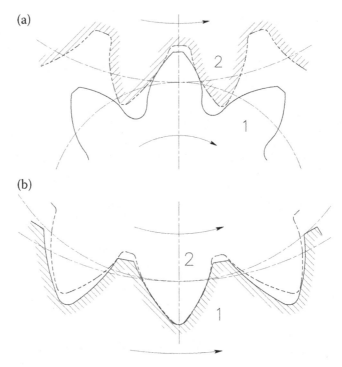

FIGURE 9.15
Shaper cutter profile generating for external (a) and internal (b) gears; 1: gear profile (solid contour); 2: shaper cutter profile (dashed contour).

$$\gamma_g = \frac{2B_g B_g'}{d_{pg}},\qquad(9.3)$$

where $B_g B_g'$ is the length of the arc $B_g B_g'$, and d_{pg} is the gear pitch diameter in the mesh with the shaper cutter.

Point A_g is in position A_g' where the gear tooth profile 1′ is tangent to the shaper cutter profile 2′. Line $A_g' B_g'$ is rotated back relative to the shaper cutter center on the angle γ_t, that is

$$\gamma_t = \gamma_g \frac{d_{pg}}{d_{pt}},\qquad(9.4)$$

where d_{pg} is the gear pitch diameter in the mesh with the shaper cutter.

This movement puts the point A_g' in position A_t at the shaper cutter profile that corresponds to the point A_g at the gear tooth profile. This approach allows for the definition of any shaper cutter profile point corresponding to the point at the gear tooth profile.

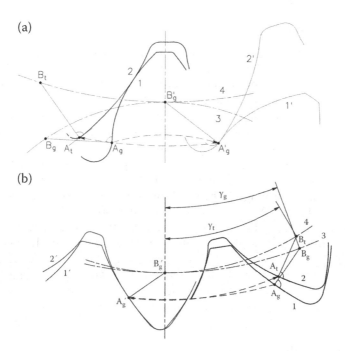

FIGURE 9.16
Generating gear shaper cutter profile definition for external (a) and internal (b) gears. 1 and 1´:
gear profile positions; 2 and 2´: shaper cutter profile positions; 3: gear pitch circle in mesh with
shaper cutter; 4: shaper cutter pitch circle in mesh with gear.

One of the benefits of the gear generating machining process is the
possibility of using one tool (hob, rack, or shaper cutter) for machining
gears with different numbers of teeth. This allows the reduction of tooling
inventory and cost of low and medium volume gear production, where a
tooling share per one gear is relatively high. In general, traditionally designed mating gears are machined with the same generating gear cutter.
However, in mass gear production gear cutting machines are typically set
up to produce one gear, and they use a set of tools, including the gear
cutters, dedicated to this particular gear.

Directly designed gears require a custom dedicated generating tool for
every gear with different numbers of teeth. This increases tooling inventory
and related gear costs when production volume is low. Although, even in
this case, the application of directly designed gears is beneficial if their
improved performance justifies some production cost increase, like, for
example, in aerospace and racing transmissions. In mass gear production, a
cost-share of custom dedicated machining tool per gear is low, and, as a
result, the cost of the directly designed gears becomes practically equal to
that of the standard ones. It makes them applicable for automotive,
agriculture, and many other industries.

9.1.1.3 Contour Machining

Contour machining (Figure 9.17) is not a highly productive gear fabrication method, but unlike form or generating machining, it does not require special tooling. It makes it very useful for gear prototyping and quick fabrication of a relatively small quantity of gears.

Contour gear machining processes include the CNC milling, wire-cut EDM, laser cutting, water jet cutting, etc. To achieve the required accuracy and surface finish the contour machining process may necessitate several passes. The final pass typically removes a tiny amount of material.

Most of these processes are suitable only for spur gears; though, the CNC ball cutter contour milling can be used for helical gears as well. The contour cutting path (Figure 9.18) is defined with the offset S from the nominal (average material condition) gear profile. This offset contains half of the cut width W_c and some additional offset that depends on the machining process. This additional offset may include the stock for final machining, overcut or overburn (for wire-cut EDM and laser machining), and a defective layer that should be removed by the following tooth surface treatment operation, for example, polishing.

9.1.2 Gear Forming

Gear forming fabrication technologies have gained popularity in the last several decades, providing a high benefit-cost ratio for mass-produced gear drives. The gear forming fabrication technologies include powder metal (PM) processing, plastic and metal injection molding, net forging, die casting, extrusion, gear and worm rolling, etc. The progress of these technologies

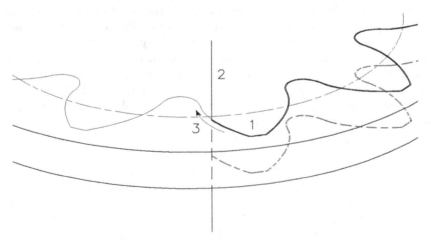

FIGURE 9.17
Schematic of contour gear machining. 1: gear profile; 2: cutting tool; 3: cutting direction.

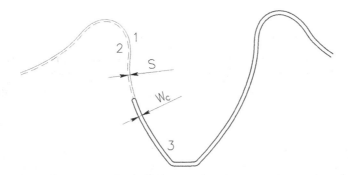

FIGURE 9.18
Contour cutting path: 1: nominal (average material condition) gear profile; 2: tool path; machined gear profile; W_c: cut width; S: tool path offset.

allowed a significant increase in their accuracy. As a result, the usage of formed gears has also increased considerably. Most of these very different gear forming fabrication processes have a similar tooling component that defines a gear shape and its accuracy – the tooling (mold or die) cavity. The gear forming process cavity is dedicated to a particular gear profile. It makes the Direct Gear Design® method very acceptable for the gear forming processes, because the production cost of the custom optimized gears, in this case, is practically the same as that of the similar size standard gears.

This section presents basic principles of the powder metal (PM) processing and plastic injection molding gear forming fabrication.

9.1.2.1 PM Gear Processing (Written by Dr. Anders Flodin)

Powder metal (PM) gears are commonly used for mass production, starting at around 10,000 parts. The high initial tooling cost makes the technology unsuitable for smaller batches. Prototype PM gears are therefore usually cut from a sintered puck using conventional cutting methods to evaluate their performance before investing in compaction tools.

Powder metal gears are compacted in a die using mechanical, hydraulic, or electric ball-screw type presses. Each of these presses has its advantages and disadvantages. Table 9.2 illustrates the differences in the applicability for gear manufacturing.

For example, planetary gearboxes of power tools utilize small PM gears with accuracy class 9–10 per ISO 1328 standard. Such gears are made on 150-ton mechanical presses, possibly with a multi-cavity tool making 2–4 gears per stroke. An electrical press could also be applicable to make power tool gears, but they are not that common on the market yet due to their novelty. A hydraulic press could do the job, but it would hurt production economics due to slower compaction speed and generally a more expensive press to run per hour.

TABLE 9.2

Differences in Applicability of Different PM Gear Presses

	Manufacturing speed	Tolerance class	Size of gears	Gear body complexity	Helix angle
Mechanical	++	+	++	+	With Alvier adapter
Electric	+++	+++	+	+	With Alvier adapter
Hydraulic	+	+++	+++	+++	With Alvier adapter

More complex helical gears are manufactured using a hydraulic press with an Alvier adaptor. The tools and drive system for the upper punch (or lower punch for internal gears) are designed based on the gear module, pitch diameter, helix angle, and face width. Figure 9.19 shows the movements of the main parts of the toolset for helical PM gears.

Figure 9.20 presents an example of such a helical automotive transmission gear. All gear features are produced by the net-shape compaction. The tooth flanks are ground after case carburization and hardening to achieve the required tolerances and lead crowning.

The number of PM gear processing steps should be minimized for cost reasons. Here are some process routes using PM gear technology ordered from simple gear to complex gear:

1. *Compaction – Sinter*. Low-cost nonhardened gears made out of cheap iron–carbon alloys with a density of about 6.6–6.8 g/cc. Typical applications are low loaded gears for non-critical machinery, lawn and garden, adjustment or control, etc.

2. *Compaction – Sinter hardening*. Sinter hardening is a hardening step built into some sinter furnaces. It cools the gears using atmosphere gas and a fan inside the end section of the sintering furnace. The cooling speed is set by fan speed, and the gas is cooled through a heat exchanger. This creates martensitic structures with increased hardness and strength as a consequence. Typical material has high admixed carbon content (0.4–0.6%) with 1–2% copper, and the rest is iron. These gears can carry more load, and due to increased hardness, they wear less than just sintered gears. Typically, used in lawn and garden and power tool gears.

3. *Compaction – Sinter – Nitriding or Induction hardening – (Hard finish)*. Gears made with this process typically need extra strength but low distortion. Some lower stressed ring gears in automatic transmissions for cars can be made with this process. The material is

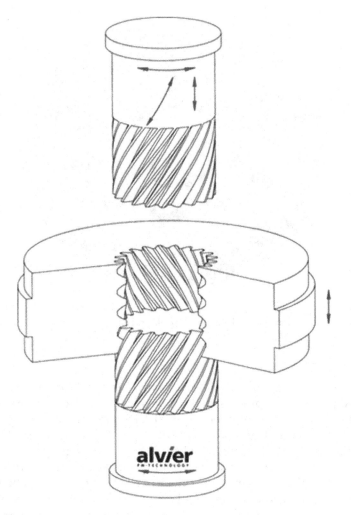

FIGURE 9.19
Basic tool concept for helical gears (Courtesy of Höganäs AB, Höganäs, Sweden. With permission.)

designed for the heat treatment process and is typically alloyed with 1.5% Mo or 0.85%Mo and 1%Ni, such as Astaloy 85Mo with an admixed nickel or Astaloy Mo, Astaloy CrA would be another alternative. The carbon content is usually 0.2–0.3%, and the density is around 7.2 g/cc in sintered condition. The nitriding process is plasma controlled for best results, and any induction hardening of PM should be dual frequency, usually with slightly (seconds) longer processing time due to the lower thermal conductivity of PM steel. Hard finishing is optional but used in some battery electric vehicles for noise reduction.

FIGURE 9.20
Automotive transmission gear with 1.9 mm module and 32° helix angle (Courtesy of Höganäs AB, Höganäs, Sweden. With permission.)

4. *Compaction – Sinter – Case carburizing – Temper – Hard finish.* Gears made with this process path are typically high-end gears. They usually are hard machined to reach tolerance class 6–7 per the ISO 1328 standard. This process results in tooth root bending fatigue limits and pitting limits equal to 75% of a wrought gear that has been case-hardened and ground to the same tolerances. To match the wrought steel gear performance, a densification step between the sinter and case carburizing is required. Typical materials would be Astaloy Mo, Astaloy 85Mo for smaller module gears, or Astaloy CrA with 1–2%Ni. The carbon content should be kept at around 0.2% to get the required compressive residual stresses from the BCC to FCC transformation and absorption of carbon atoms. The base density in these gears is usually 7.2 g/cc or higher. After the densification, the surface layer density is practically the same as in solid steel (~7.8 g/cc). These process steps are used in highly loaded gear drives, for example, for automotive transmissions.

Figure 9.21 presents an experimental asymmetric gear stage of a Saab 9-5 that was used as a demonstrator [92]. The driving shaft pinion was made conventionally out of solid steel and the driven gear out of the powder metal alloy produced by Höganäs AB.

FIGURE 9.21
Asymmetric tooth gear pair; 1: machined and ground pinion, 2: powder metal gear with ground teeth (Courtesy of Höganäs AB, Höganäs, Sweden. With permission.)

Several more processes can be used for the manufacturing of PM gears, such as powder forging, quench and temper hardening, vacuum carburizing, shot peening, hot isostatic pressing (HIP), and more. Generally, most of these processes and the process equipment used for wrought gears should also work for powder metal gears with some tweaking of the process parameters, usually with shorter processing time due to faster diffusion of process gases into the lattice of the steel.

An additional reserve of increasing load capacity of powder metal gears is the tooth profile densification that increases gear tooth strength and wear resistance. This secondary operation allows achieving powder metal gear performance that is close to that of solid gear steels [93]. Densification of PM gears is a collection of methods to locally increase the density in the high stress areas of the gear profile, i.e., at the tooth root and flank. Roll densification is the most common method; it increases density on the flank surface and in the high stress area at the tooth root. It also shapes the flanks to create lead crowning and compensate for heat treatment distortions. The PM gear is rolled between two master gears that are moving closer together during the 9–10 s process (Figure 9.22), achieving the 7–8 ISO 1328 tolerance class after case hardening [94].

Figure 9.23 shows the densified tooth flank and root surfaces of a car transmission gear. In this case, there was no hard tooth machining, only roll

FIGURE 9.22
Surface densification of the PM gear between two master gears (Courtesy of Profiroll Technologies, GMBH, Germany. With permission.)

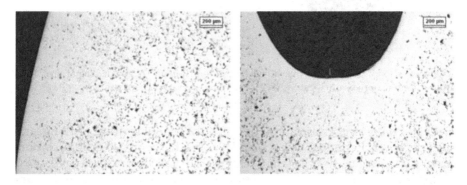

FIGURE 9.23
Densified flank (left) and root (right) of the pinion in a six-speed manual car transmission (Courtesy of Höganäs AB, Höganäs, Sweden. With permission.)

densification as the final operation. This gear had survived durability testing and was quieter when run against a ground PM gear than the wrought steel gears from the OEM. The gear material was Astaloy 85Mo. Astaloy Mo and Astaloy CrA also work well with the densification process as long as the carbon content is below 0.25%, which is typical for case hardened PM gears.

The selection of powder metal material depends on the strength requirements for the particular gear application. The strength of PM alloy is predominantly determined by the density and heat treatment. Sinter hardening, nitriding, or case carburizing sets the alloying composition to respond to the heat treatment process. A gear subjected to sinter hardening has a different alloying composition than a gear material for case hardening. The density is more dependent on the lubricant that is admixed in the powder to reduce intra-particle friction and friction between powder and tool die during compaction and ejection of the gear. The lubricant also influences productivity since it affects how fast the powder flows into the tool cavity. For example, a compaction speed of 7 gears per minute instead of 6 represents a 16% increase in productivity, which often is the result of using a better lubricant. A better lubricant also gives longer service life for the tooling and better surface finish of the part. Generally, the tool die should last for 500,000 gears, but some live for 1,000,000 or more. The punches can be refurbished every 100,000–200,000 gears. A good solid lubricant in the powder extends the service life of the tools as well and contributes to the production economics.

The wire EDM of the prototype PM gears is not the best choice compared to conventional gear machining. It tends to decarburize the surface creating micro-cracks that result in premature failures. Gear cutting works better for PM gear prototyping. However, excessive tool wear is possible due to the intermittency of porous material cutting.

9.1.2.2 Plastic Gear Molding

A plastic gear tooling cavity has a profile similar to the gear profile but adjusted for shrinkage and warpage (Figure 9.24). This adjustment predominantly affects gear size and shape accuracy. It made a proper prediction of shrinkage and warpage critical for all gear forming technologies, particularly for the plastic injection molding process. Plastic gears often have intricate body shapes, including ribs or spokes, to maintain limited

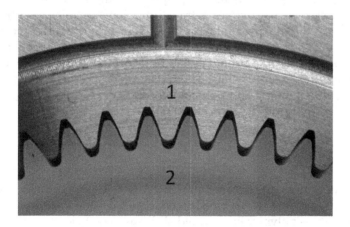

FIGURE 9.24
Plastic gear injection molding; 1 – tooling cavity, 2 – gear. (Kleiss, R.E. How to achieve a successful molded gear transmission. *Gear Technology*, July/August 2006, 42–47. With permission.)

maximum material thickness to exclude voids and for weight and cost reduction. They also can be incorporated as one piece with other mechanical components like, for example, shafts, cams, etc. These design specifics, combined with an enormous variety of available gear polymers and enhancing additives (for increased strength, thermal resistance, lubricity, etc.), make a prediction of gear shrinkage and warpage an extremely challenging task. In many cases, gear molders utilize a trial and error method with a different degree of success. Typically, this "educated guess" method works better for gears with relatively simple body shape (like, for example, the flat uniform disk with a small central hole) made out of generic unfilled polymers.

R.E. Kleiss [95] stated that "plastic does not shrink from the cavity in an isotropic fashion" and suggested using different shrinkage factors for main gear dimensions, including the tooth tip, root, and base circle diameters, and base tooth thickness. S.F. Walsh [96] analyzed the effects of material crystallinity, orientation, and cooling stress relaxation for shrinkage and warpage prediction of injection molded components. The Autodesk® Moldflow® plastic injection molding simulation software [97] predicts "part shrinkage based on processing parameters and grade-specific material data." Realization of these and similar approaches requires knowledge of specific data about polymer material grade, tooling design (number of injection gates, their size, and location, cooling system, etc.), and molding process parameters that are not always accurate and available.

An entirely different approach to molding distortion compensation was proposed by Dr. Yuriy Shekhtman [98]. His method is based on a mathematical prediction that defines a transformation function describing relations between the initially molded sample gear profile and its actual initial cavity profile. Once this function is defined, the target gear profile replaces the first molded sample profile as the transformation function variable to calculate the final cavity profile. The transformation function is based on a system of trigonometric and polynomial functions.

The initial cavity profile coordinates are

$$M_1 = K_{sh} \times D, \tag{9.5}$$

where

D – target gear profile data set, presented as X, Y – – coordinates points of the 2D CAD model typically constructed for average material conditions; K_{sh} – polymer linear mold shrinkage coefficient provided by the material supplier.

These initial cavity profile coordinates M_1 can be also presented as

$$M_1 = f(P), \tag{9.6}$$

where

P – initial sample gear profile data set, presented as X, Y – – coordinates points provided by the CMM (Coordinate Measuring Machine) inspection of actual molded gear; f – transformation function describing relations between the initial cavity and initial sample gear profiles.

Then the final cavity profile coordinates are

$$M_2 = f(D). \tag{9.7}$$

Unlike previously mentioned approaches, Dr. Shekhtman's method is based on the "black box" concept and uses only gear and cavity inspection results and math that defines the transformation function between them. It does not require knowledge of any specific data related to the polymer material, tooling, and molding process parameters.

Practical application of this method takes 8 steps:

1. *Target gear profile definition (data file #1).* The X, Y-coordinate points of the gear CAD model present a desired nominal gear profile at the average material condition. A number of these coordinate points is typically several hundred per one gear tooth.

2. *Initial cavity profile definition.* The initial cavity profile is the scaled-up target gear CAD profile using the polymer linear mold shrinkage coefficient K_{sh} from its specification.

3. *Fabrication and inspection of the initial mold cavity (data file #2).* CMM inspection produces the X, Y-coordinate points (several hundred per one tooth space) accurately describing the initial cavity.

4. *Molding Process Optimization.* Gears are molded using the initial cavity, without concern about gear shape. A goal here is to achieve a stable and repeatable molding process with the part dimensional variation significantly lower than the required accuracy tolerances. Any material flaws like voids are not acceptable. Once this goal is reached, the molding process must be "locked-in" and certified; no changes to the process are now allowed. Several dozen gears are molded using the optimized process.

5. *Representative gear specimen selection.* All molded gears are roll tested, and inspection data are analyzed. Then the most representative preliminary gear specimen is selected. This specimen should have average statistical tooth-to-tooth and total composite errors (TTE and TCE).

6. *Gear specimen inspection (data file #3).* CMM inspection produces the X, Y-coordinate points (several hundred per one gear tooth) accurately describing the initial cavity. Inspection data of the initial cavity and the gear specimen must have the same axes orientation to provide each gear tooth and its cavity space accordance.

7. *Final cavity profile definition and fabrication.* The Genetic Molding Solution software uses the gear specimen and initial cavity data (files #2 and #3) to generate a transformation function f. The target gear data (file #1) is then used as the variable of this transformation function to define the final cavity profile – the output data set. The same axis orientation of all three data files is critical. Any angular rotation or mirroring of the data points compromises the mold cavity adjustment results. The final cavity is then manufactured and given a CMM check-inspection.

8. *Final gear profile.* At last, gears are molded using the final mold cavity. The CMM data of the molded gears should be identical to the specified gear profile, within the molding process accuracy variation.

For the successful application of this method, the initial and final gear molding must be done with the same batch polymer on the same molding press using the same tool. The software uses the 2D data sets and works well for spur plastic gears with relatively low face width. For helical and spur gears with large face width, this method should use for several (typically 2–3) gear sections. Figure 9.25a shows an example of this method application for the camshaft gear. This gear is not particularly molding friendly: it has a metal over-molded shaft, two cams, six spokes, and three injecting gates located in the middle of these spokes. Mold cavity (Figure 9.25b) development for this gear using traditional methods requires considerable time and guesswork, and several mold cavity iterations. The suggested mathematical prediction method develops the desired cavity in a short time by direct calculation, with only one extra (initial) cavity.

Figure 9.25c shows a comparison of roll test graphs on the initial most representative gear specimen with the final gear sample. The initial gear roll test measurements (TTE and TCE) show insufficient accuracy. But the final gear roll test results are well inside the required TTE and TCE tolerance limits.

Dr. Shekhtman's method significantly accelerates the injection of plastic mold cavity development. It eliminates a "guess" component of the final cavity prediction and provides its profile definition by the use of direct calculation. It applies not only to plastic molded gears but also to other plastic components.

9.1.3 Additive Gear Technologies (Written by Dr. Anders Flodin)

3D printing or additive manufacturing (AM) is a relatively new technology for fabrication, not only gears but parts in general. The first functional 3D printing machines were developed during the 1980s, and today plastic filament printers are prevalent in the market. The advantage of AM is that theoretically, no tooling is required, and the capability to produce some

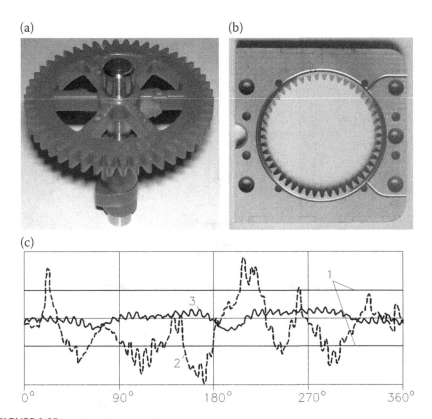

FIGURE 9.25

(a): cam shaft gear, (b): gear mold cavity, (c): roll test chart overlay, 1: total composite tolerance limits, 2: initial specimen chart (solid contour), 3: final gear chart (dashed contour). (From Kapelevich, A.L., et al., Turning an art into science. *Motion System Design*, August 2005, 26–31. With permission.)

designs that would be impossible to manufacture with conventional methods. This section is about the essentials of the 3D gear printing technology and also describes the characteristics of the layer by layer process that is the most common to produce both plastic and steel gear prints [99].

There are several 3D gear printing methods:

1. *Fused Filament Deposition (FFD)*. The most common way to print plastic gears is the filament method. A plastic filament on a spool is fed into a heated extruder head that extrudes the molten plastic in a thin line, making layers on top of one another to build the part from bottom to top. A principle of the FFD method is shown in Figure 9.26. The layer thickness is usually 0.1 mm, but this is adjustable in the slicing software. These types of printers are

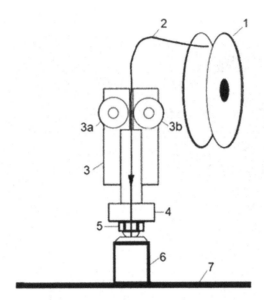

FIGURE 9.26
Principle of FFD method. 1: spool; 2: filament; 3: feeder with feeder rollers (3a and 3b); 4: heating element; 5: extruder nozzle; 6: printed part/support; 7: heated bedplate. (Courtesy of Höganäs AB, Höganäs, Sweden. With permission.)

commonly available, and many AM enthusiasts have them in their homes. Typical materials are acrylonitrile butadiene styrene (ABS) and polylactic acid (PLA) polymers. The layers are clearly visible, and the surface roughness perpendicular to the layers is higher than along the layers. For a gear, this means that the surface roughness is better in the sliding direction on the flanks of a cylindrical gear compared to the perpendicular one. The gear accuracy is not high but sufficient for functioning plastic gear trains. The weaknesses of these prints are the tolerances (around 0.1 mm), and the intra layer adhesion, meaning the parts don´t exhibit the same mechanical properties in all directions. They tend to break between layers rather than perpendicular to them.

The FFD method for metal parts utilizes material composition similar to that for metal injection molding (MIM), where the binder is a polymer-wax with a high percentage of metal powder that is alloyed to work with the consecutive sintering and hardening processes. The advantage of this process is a low investment cost, but the drawbacks are similar to MIM fabrication, such as slow de-binding and significant shrinkage of the gears during the sintering process that follows after printing. This technology can create hollow structures and requires no machining to remove gears from the bedplate. The resulting gear accuracy is about class 12 per ISO 1328 standard, and secondary machining operation is necessary to achieve higher gear accuracy.

2. *Stereo Lithography*. This is one of the first methods to be commercialized for more professional prints. The equipment is significantly more expensive than the filament (FFD) method, and it is also a bit more complex to use. UV light is directed into the bath of a photopolymer liquid and draws the part, layer by layer. The UV light cures the liquid. It offers higher resolution printing than many other 3D printing technologies, allowing users to print parts with fine details and surface finishes. Smaller low-cost machines are now available, which makes this technology more attractive.

3. *Selective Laser Sintering (SLS)*. This is a 3D printing technology that produces highly accurate and durable parts. It is similar to the LPBF process for steel (see method #6, described below), but instead of fusing steel particles, it fuses polymer powder particles. The parts require cleaning after printing since it is all in a powder bed. Intricate or hollow shapes have to be designed with an opening so that plastic powder trapped inside, for instance, can be removed by compressed air or brushing. Cleaning is often by manual labor even though there is ongoing work to automate it for series production. Plus, the removed material can be contaminated and should be discarded, which increases the part cost.

4. *Laser and Ion Beam Methods*. These two beam methods work with a powder bed where the particles are sinter-bonded by the beam through local temperature elevation/melting. Several beams can work simultaneously to speed up the process. However, too much energy into the bed creates distortions. It is a layered process meaning the parts are built up layer by layer. Layer thickness is typically 10–50 μm. The thinner the layers, the more printing time is required. Figure 9.27 shows the layer build-up principle and typical print parameters.

Figure 9.28 shows the principle of Laser printing. Each layer of powder is distributed with a roller or scraper. The layer adhesion is monitored to avoid insufficient bonding and anisotropic properties. There is a balance between the printing parameters such as the focal diameter, speed, hatching distance all contribute to surface roughness, temperature build, distortions, phase transformations in the steel, particle bonding, productivity, etc.

Figure 9.29 shows a build plate after Laser Powder Bed Fusion (LPBF) printing. After the print, the whole build plate should be annealed to

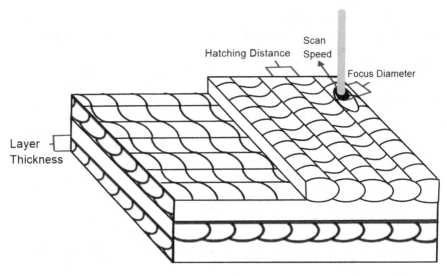

FIGURE 9.27
Typical print parameters: layer thickness – –20 μm, scan speed – –1100 mm/s, hatching distance – –0.09 mm. (Courtesy of Höganäs AB, Höganäs, Sweden. With permission.)

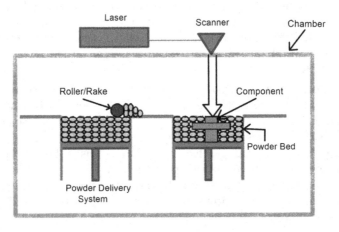

FIGURE 9.28
Principle of laser and ion beam printing. (Courtesy of Höganäs AB, Höganäs, Sweden. With permission.)

release the residual stress that has been introduced by the heat input from the laser or ion beam. The parts should be cut off from the build plate, often with a band saw, and the parts' faces turned since the sawed surface is rough. The resulting gear accuracy is about class 12 per ISO 1328 standard, and secondary machining operation of the tooth flanks and the

FIGURE 9.29
Printed build plate; printing time – 24 hours, smaller gear module – –1.59 mm, outer diameter – 32 mm (Courtesy of Höganäs AB, Höganäs, Sweden. With permission.)

bore is necessary to achieve required gear accuracy. Most materials available for printing are high alloys such as stainless steel, titanium, and tool steel.

5. *Binder Jet.* The binder jet (BJ) process has some features that set it apart from the other methods described above. It is still a powder bed type printing process, but instead of a laser fusing, a binder or glue is sprayed on the particles where they are supposed to bond, similar to an inkjet printer head that is sprinkling ink on a paper. The BJ printing utilizes very fine powder that allows printing of intricate parts with much smaller details than what is possible with lasers or filaments. However, the build volume is less than what is typical for laser and filament printers, and the part size is usually less than 50 mm. After the printing and the cleaning, which can be automated, the parts are sintered to reach the required strength and geometry, causing some shrinkage. Surface roughness is around Ra 6 μm after printing, and depending on the material, the ultimate tensile strength is from 550 MPa for 316 L to 1200 MPa for the highly alloyed DM427 material.

Additive gear processing is relatively new, rapidly developing gear fabrication technology, ideally suitable for manufacturing gears, defined by the Direct Gear Design® method.

9.2 Gear Measurement

Inspection is a critical stage of the gear production process. It is also essential for the development of new gear transmissions. A comprehensive gear inspection before assembly and prototype testing allow isolating potential design issues from manufacturing errors. It makes it possible to draw correct conclusions based on the prototype testing results.

Manufacturing of directly designed gears not only requires custom tooling but also affects gear measurement. This chapter presents definitions of main inspection dimensions and parameters for the directly designed spur and helical, external, and internal gears with symmetric and asymmetric teeth. All equations of this section are for the asymmetric tooth gears as the most general form of directly designed involute gears. However, they are suitable for the symmetric tooth gears that have identical parameters of the opposite tooth flanks.

9.2.1 Measurement over (between) Balls or Pins

Measurement over (for gears with external teeth) or between (for gears with internal teeth) balls or pins is an indirect way to inspect the tooth thickness at the given reference diameter.

9.2.1.1 Measurement of Spur Gears

The position of the measuring ball or pin center is shown in Figure 9.30. The involute angles α_{gd} and α_{gc} at the ball or pin center location circle diameter d_g are defined by [100, 101]:

For the external gear:

$$inv(\alpha_{gd}) + inv(\alpha_{gc}) = inv(\nu_d) + inv(\nu_c) + \frac{D}{d_{bd}} + \frac{D}{d_{bc}} - \frac{2\pi}{z}. \tag{9.8}$$

For the internal gear:

$$inv(\alpha_{gd}) + inv(\alpha_{gc}) = inv(\nu_d) + inv(\nu_c) - \frac{D}{d_{bd}} - \frac{D}{d_{bc}}. \tag{9.9}$$

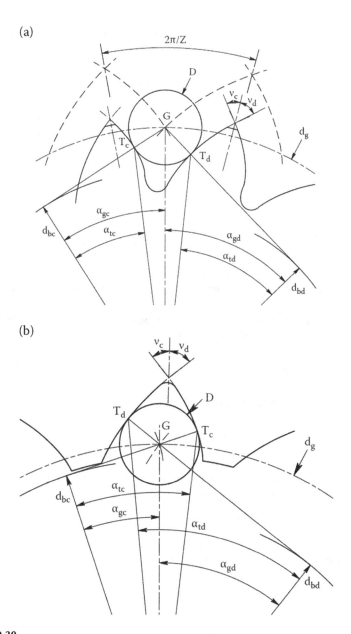

FIGURE 9.30
Ball or pin position: (a) external gear, (b) internal gear. D – ball or pin diameter; G – center of ball or pin. (From Kapelevich A.L., Measurement of directly designed gears with symmetric and asymmetric teeth. *Gear Technology*, January/February 2011, 60–65. With permission.)

The ball or pin center location circle diameter d_g is

$$d_g = \frac{d_{bd}}{\cos \alpha_{gd}} = \frac{d_{bc}}{\cos \alpha_{gc}}. \tag{9.10}$$

The involute angles α_{td} and α_{tc} in the ball or pin contact points T_d and T_c are

$$\alpha_{td} = \arctan\left(\tan \alpha_{gd} \mp \frac{D}{d_{bd}} \right) \tag{9.11}$$

and

$$\alpha_{tc} = \arctan\left(\tan \alpha_{gc} \mp \frac{D}{d_{bc}} \right), \tag{9.12}$$

where the sign "−" is for an external gear and the sign "+" is for an internal gear.

Then a measurement over two balls or pins for an external gear is:

For even number of teeth (Figure 9.31a):

$$M = d_g + D. \tag{9.13}$$

For odd number of teeth (Figure 9.31b):

$$M = d_g \cdot \cos \frac{\pi}{2z} + D. \tag{9.14}$$

A measurement between two balls or pins for an internal gear is:

For even number of teeth (Figure 9.32a):

$$M = d_g - D. \tag{9.15}$$

For odd number of teeth (Figure 9.32b):

$$M = d_g \cdot \cos \frac{\pi}{2z} - D. \tag{9.16}$$

The ball or pin contact points T_d and T_c should always be located on the involute flanks. They must not contact the tooth profile at the tooth tip radius or chamfer and at the root fillet profile (Figure 9.33). These conditions are described by the contact point involute angle limits:

For external gears:

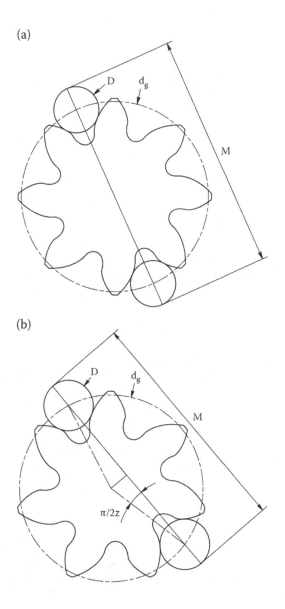

FIGURE 9.31
Measurement over balls or pins for external spur gears: (a) even number of teeth, (b) odd number of teeth. (From Kapelevich A.L., Measurement of directly designed gears with symmetric and asymmetric teeth. *Gear Technology*, January/February 2011, 60–65. With permission.)

$$\arccos\frac{d_{bd}}{d_{fd}} < \alpha_{td} < \arccos\frac{d_{bd}}{d_a} \text{ and } \arccos\frac{d_{bc}}{d_{fc}} < \alpha_{tc} < \arccos\frac{d_{bc}}{d_a}. \quad (9.17)$$

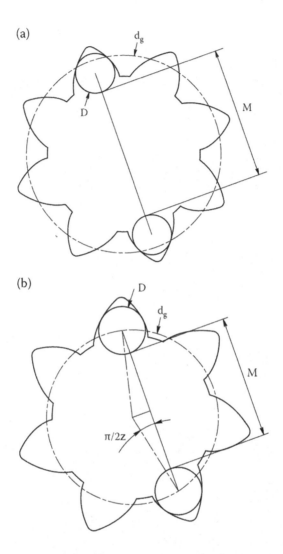

FIGURE 9.32
Measurement over balls or pins for internal spur gears: (a) even number of teeth, (b) odd number of teeth. (From Kapelevich A.L., Measurement of directly designed gears with symmetric and asymmetric teeth. *Gear Technology*, January/February 2011, 60–65. With permission.)

For internal gears:

$$\arccos\frac{d_{bd}}{d_a} < \alpha_{td} < \arccos\frac{d_{bd}}{d_{fd}} \text{ and } \arccos\frac{d_{bc}}{d_a} < \alpha_{tc} < \arccos\frac{d_{bc}}{d_{fc}}, \quad (9.18)$$

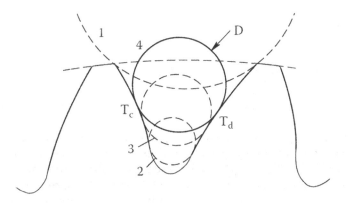

FIGURE 9.33
Ball or pin diameter selection. 1: diameter is too large resulting in a contact point at the tooth tip corner; 2: diameter is too small resulting in a contact point at the tooth root fillet; 3: diameter is acceptable, but a small tip caliper (or micrometer) is required to fit between gear teeth; 4: suitable ball or pin diameter.

where d_a is the tooth tip diameter, and d_{fd} and d_{fc} are the drive and coast tooth flank form diameters.

For measurement convenience the ball or pin surface should be above the gear tooth tips; otherwise the caliper (or micrometer) should have small tips to fit between gear teeth. This condition can be presented as

$$D > |d_a - d_g|. \tag{9.19}$$

For external gears this inequality should be solved with Equations (9.8) and (9.10), and for internal gear with Equations (9.9) and (9.10), to define a proper ball or pin diameter.

9.2.1.2 Measurement of Helical Gears

The involute angles α_{gd} and α_{gc} at the ball or pin location circle diameter d_g are defined by (9.11):

For an external gear:

$$inv(\alpha_{gd}) + inv(\alpha_{gc}) = inv(v_d) + inv(v_c) + \frac{D}{d_{bd} \times \cos\beta_{bd}} + \frac{D}{d_{bc} \times \cos\beta_{bc}} - \frac{2\pi}{z}. \tag{9.20}$$

For an internal gear:

$$inv(\alpha_{gd}) + inv(\alpha_{gc}) = inv(v_d) + inv(v_c) - \frac{D}{d_{bd} \times \cos\beta_{bd}} - \frac{D}{d_{bc} \times \cos\beta_{bc}}, \tag{9.21}$$

where β_{bd} and β_{bc} – helix angles at the drive and coast base diameters that are equal to

$$\beta_{bd} = \arctan(\tan\beta \times \cos\alpha_d), \tag{9.22}$$

$$\beta_{bc} = \arctan(\tan\beta \times \cos\alpha_c). \tag{9.23}$$

Then the ball or pin center location circle diameter d_g is defined by (9.10). This equation is for the external helical gear pin center location circle definition. Cylindrical pins cannot be used to measure the internal helical gears because the pin surface cannot be tangent to the internal helical gear tooth flanks. They should be inspected by a measurement between balls. The ball or pin diameters should also satisfy conditions (9.17) – (9.19). When diameter d_g is known, measurements over two balls for external helical gears (Figure 9.34) and between two balls for internal helical gears (Figure 9.35) are defined by Equations (9.13) and (9.14), and (9.15) and (9.13), accordingly.

Measurement over two pins for external helical gears with an even number of teeth is also defined by Equation (9.13). Measurement over two pins for external helical gears with an odd number of teeth is considered inappropriate by some gear experts because the shortest distance between

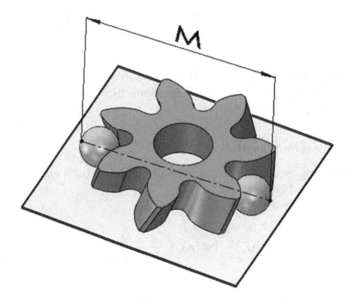

FIGURE 9.34
Measurement over balls of external helical gear. (From Kapelevich A.L., Measurement of directly designed gears with symmetric and asymmetric teeth. *Gear Technology*, January/ February 2011, 60–65. With permission.)

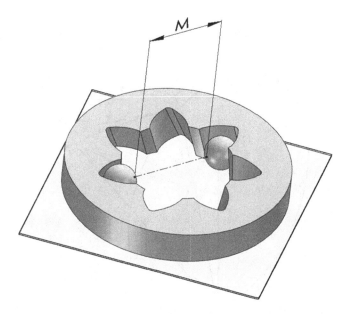

FIGURE 9.35
Measurement between balls of internal helical gear. (From Kapelevich A.L., Measurement of directly designed gears with symmetric and asymmetric teeth. *Gear Technology*, January/February 2011, 60–65. With permission.)

the pin centers does not intersect the gear axis. However, this type of measurement is commonly used in gear production, and it is necessary to provide a correct definition of measurement over two pins for external helical gears with an odd number of teeth.

For external helical gears with an odd number of teeth, the shortest distance L between the pin centers does not lie in the transverse section of the circle diameter d_p (Figure 9.36). This distance definition is described in [102] as

$$L = \frac{d_g}{2 \times \tan \beta_g} \sqrt{\lambda^2 + 4 \times \left(\tan \beta_g \times \cos \left(\frac{\pi}{2z} + \frac{\lambda}{2} \right) \right)^2}, \qquad (9.24)$$

where the helix angle at the pin center diameter β_g is

$$\beta_g = \arctan \left(\frac{\tan \beta_{bd}}{\cos \alpha_{gd}} \right) = \arctan \left(\frac{\tan \beta_{bc}}{\cos \alpha_{gc}} \right). \qquad (9.25)$$

and the angle λ is a solution of the equation

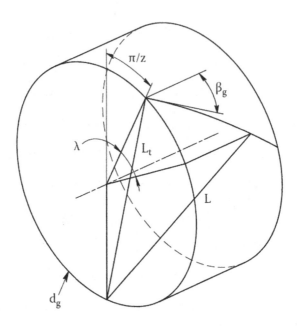

FIGURE 9.36
Definition of the distance between the pin centers for the helical gears with an odd number of teeth. (From Kapelevich A.L., Measurement of directly designed gears with symmetric and asymmetric teeth. *Gear Technology*, January/February 2011, 60–65. With permission.)

$$\frac{\lambda}{\tan \beta_g} - \sin\left(\frac{\pi}{z} + \lambda\right) = 0. \tag{9.26}$$

Then the measurement over two pins for external helical gears with an odd number of teeth (Figure 9.37) is

$$M = L + D. \tag{9.27}$$

9.2.2 Span Measurement

Span measurement is another way to inspect the tooth thickness at a given reference diameter for spur gears with external and internal symmetric teeth, and for helical gears with external symmetric teeth. This inspection method is impractical for helical gears with internal symmetric teeth. It cannot be applied for gears with asymmetric teeth because it is impossible to have a mutual tangent line to two concentric base cylinders of the drive and coast asymmetric tooth flanks. Span measurement is a distance over (for external tooth gears) or between (for

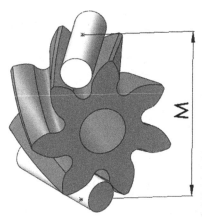

FIGURE 9.37
Measurement over pins of the external helical gear with an odd number of teeth. (From Kapelevich A.L., Measurement of directly designed gears with symmetric and asymmetric teeth. *Gear Technology*, January/February 2011, 60–65. With permission.)

internal tooth gears) several teeth, along the tangent line to the base cylinder (Figure 9.38).

For external spur and helical gears (Figures 9.38a and 9.38b), the span measurement over z_w teeth is

$$W = (S_b + (z_w - 1) \times p_b) \times \cos \beta_b, \tag{9.28}$$

where S_b is the tooth thickness at the base diameter

$$S_b = S \times \cos \alpha + d_b \times inv(\alpha), \tag{9.29}$$

S and α are gear tooth thickness and involute profile angle at the reference diameter d, p_b is the circular pitch at the base diameter

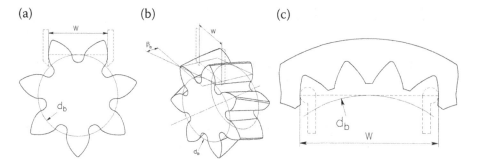

FIGURE 9.38
Span measurement: (a) external spur gear, (b) external helical gear, (c) internal spur gear. ((a and b) From Kapelevich A.L., Measurement of directly designed gears with symmetric and asymmetric teeth. *Gear Technology*, January/February 2011, 60–65. With permission.)

$$p_b = \frac{\pi \times d_b}{z},$$ (9.30)

β_b is the helix angle at the base diameter

$$\beta_b = \arctan(\tan \beta \times \cos \alpha),$$ (9.31)

z_w is number of teeth for span measurement

$$2 \le z_w \le z_{w\,max}.$$ (9.32)

For external gears a maximum number of the spanned teeth z_{wmax} is

$$z_{w\,max} = \frac{\sqrt{d_a^2 - d_b^2} - S_b}{p_b}.$$ (9.33)

For internal spur gears (Figure 9.38c), the span measurement between z_w teeth is

$$W = (Sp_b + (z_w - 2) \times p_b),$$ (9.34)

where Sp_b is the tooth space width at the base diameter:
   for the Metric system gears:

$$Sp_b = (\pi \times m - S) \times \cos \alpha + d_b \times inv(\alpha),$$ (9.35)

for the English system gears:

$$Sp_b = (\pi / DP - S) \times \cos \alpha + d_b \times inv(\alpha).$$ (9.36)

z_w is number of teeth for span measurement

$$3 \le z_w \le z_{w\,max},$$ (9.37)

For internal spur gears a maximum number of teeth for span measurement z_{wmax} is

$$z_{w\,max} = \frac{\sqrt{d_f^2 - d_b^2} - Sp_b}{p_b}.$$ (9.38)

Calipers, micrometers, or special gages are used for span measurement.

9.2.3 Composite Gear Inspection

There are two types of composite gear inspection: single- and double-flank composite testing [103]. Single-flank composite testing is used for the mating gears at a fixed center distance for transmission error component measurement that includes adjacent pitch variation, total accumulated pitch variation, tooth-to-tooth transmission variation, and total transmission variation. This type of testing applied to custom directly designed gears is practically the same as for conventional gears, except for the gears with asymmetric teeth that use both flanks for torque or motion transmission. In this case, opposite flanks require separate testing. This method provides a good indication of gear pair functionality because it checks two mating gears.

Double-flank composite testing has the inspected gear mounted on a rolling fixture (roll tester) with a tight spring-loaded mesh with a master gear. Deviations of the center distance during gear rotation indicate the tooth-to-tooth composite error (TTE) and total composite error (TCE). Modern roll testers with a computerized data acquisition system also allow evaluation of a tooth thickness and radial runout. Double-flank roll testing is a quick and inexpensive way to separate acceptable and defective gears. However, it does not indicate which gear dimension or accuracy parameter is responsible for excessive TTE and TCE. It also does not recognize which gear tooth flank is a major contributor to composite errors. Double-flank composite testing does not provide sufficient data about actual gear pair functionality because it checks only one gear in mesh with the master gear.

This type of testing is also applied to asymmetric gears. In this case, it requires the asymmetric master gears.

9.2.4 Elemental Gear Inspection

Elemental gear inspection utilizes coordinate measuring machines (CMM) (Figure 9.39). It allows mapping a surface of all teeth, including the fillet profiles. It provides measurement results for involute flank elemental accuracy parameters: radial runout tolerance, pitch variation, profile tolerance, and lead or tooth alignment tolerance. Directly designed gears have an optimized tooth profile, including the root fillet. Therefore, the actual accuracy of the root fillet profile should be also inspected by a CMM. The gear tooth (including the root fillet) CAD profile at the average material condition is used for the CMM inspection. The data set also includes the involute flank and fillet profile tolerances that are established, depending on required gear accuracy and also manufacturing technology. The CMM is programmed to indicate if

FIGURE 9.39
CMM measurement of asymmetric gear. (From A.L. Kapelevich. Measurement of directly designed gears with symmetric and asymmetric teeth. *Gear Technology*, January/February 2011, 60–65. With permission.)

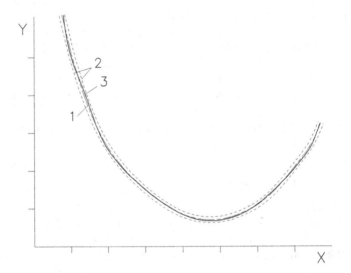

FIGURE 9.40
CMM inspection chart of the asymmetric gear root fillet profile. 1 – nominal (average material condition) profile (dash-dotted contour); 2 – tolerance corridor (dashed contour); 3 – actual inspected profile (solid thick contour).

the inspected tooth profile points lie within the corridor defined by the CAD tooth profile ± a profile tolerance. Figure 9.40 shows an example of the CMM inspection chart of the optimized root fillet profile of the asymmetric tooth.

10

Traditional vs Direct Gear Design®

The benefits of the traditional gear design are well known: it is comprehensively standardized, supported by standard design software, tooling, and a massive volume of experimental data. It provides acceptable solutions for practically all types of gear drive applications. However, a major weakness is its limited capability to optimize gear tooth geometry. This chapter is not intended to undermine or criticize the traditional gear design approach, but to compare it to Direct Gear Design®, helping a gear engineer to choose the most suitable design method for a particular gear application.

10.1 Comparable Geometry and Stress Analysis

In many cases, advanced gear geometry performance results are compared with the standard 20° pressure angle gear performance, indicating impressive advantages of new gear geometry. Although such comparison makes some sense, because the baseline performance of the standard gears is well known, it cannot be considered fully fair. Standard gears are universally applicable, but they by no means are suitable for high performance demanding applications. New advanced performance gear geometry should be compared to the best currently existing solutions for specific applications to evaluate its true benefits. Such cutting-edge baseline gear geometry can be found, for example, in modern aerospace gear transmissions. Although this comparison approach does not usually result in very impressive numbers, it puts this new gear geometry side by side with the best existing ones to assess its true benefits. Table 10.1 presents directly designed optimized gears with symmetric and asymmetric teeth, in comparison with the traditionally designed aerospace type gears that have a high 25° pressure angle and a full circular root fillet.

Table 10.2 presents a comparison of the directly designed symmetric and asymmetric gears to the traditionally designed gears that have high (>2.0) transverse contact ratio and a full circular root fillet. Such gears are also used in the aerospace gear transmissions [104].

TABLE 10.1

Comparison of Directly and Traditionally Designed Conventional Gears

Design method	Traditional (baseline)		Direct			
Tooth shape	Symmetric		Symmetric		Asymmetric	
Gear mesh	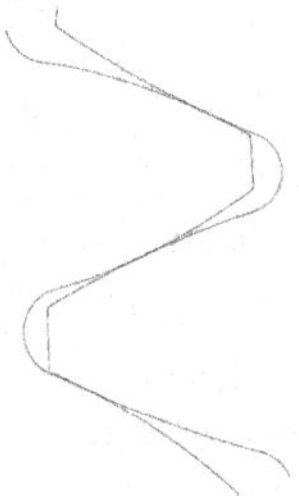		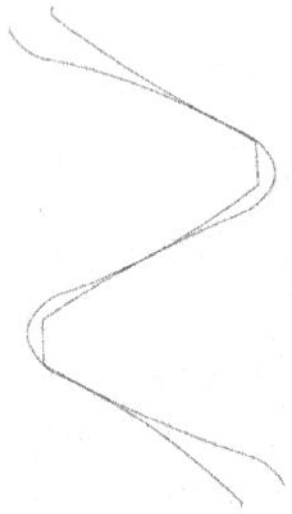			
Gears	Pinion	Gear	Pinion	Gear	Pinion	Gear
Number of Teeth	27	49	27	49	27	49
Module, mm	3.0		3.0		3.0	
X-shift	0.09	−0.09	N/A	N/A	N/A	N/A
Root fillet profile	Full circle		Optimized			
Operating pressure angle	25°		27°		32°/18°*	
Drive contact ratio	1.49		1.49		1.49/1.98*	
Pitch diameter, mm	81.0	147.0	81.0	147.0	81.0	147.0
Outer diameter, mm	87.540	152.46	87.444	153.133	87.895	153.645
Root diameter, mm	74.285	138.962	74.185	139.80	73.685	139.488
Root clearance, mm	0.628	0.749	0.341	0.380	0.335	0.308
Tooth thickness at pitch diameter, mm	4.955	4.469	4.873	4.551	4.873	4.551
Tooth thickness at outer diameter, mm	1.543	1.797	1.248	1.244	1.13	1.13
Center distance, mm	114.0		114.0		114.0	
Face width, mm	30	30	30	30	30	30

(Continued)

TABLE 10.1 (Continued)

Design method	Traditional (baseline)		Direct			
Tooth shape	Symmetric		Symmetric		Asymmetric	
Gear mesh						
Gears	Pinion	Gear	Pinion	Gear	Pinion	Gear
Driving torque, Nm	300		300		300	
Bending stress, MPa	210	213	178(−15%)	179(−16%)	182(−13%)	183(−14%)
Contact stress, MPa		958	937(−2%)		886(−7.5%)	
Drive flank specific sliding velocity	0.258	0.230	0.241	0.241	0.228	0.228
Bearing load, N		8172	8313(+2%)		8734(+7%)	
Maximum tooth tip deflection, mm	0.0054	0.0049	0.0047	0.0048	0.0071	0.0058

Note
* Drive/coast flank

10.2 Gear Testing Results Comparison

The Rotorcraft division of the Boeing Company has experimentally compared the traditional and Direct Gear Design® approaches [105]. Directly designed symmetric and asymmetric tooth gears were analyzed to determine their root bending and contact stresses relative to symmetric involute gear tooth form, which is representative of helicopter main drive gears. Three types of gear test specimens were designed, fabricated, and tested. The first type is the baseline traditionally designed symmetric tooth gears. The second type is directly designed symmetric gears, similar to the baseline gears, but with an optimized root fillet. The third type is asymmetric tooth gears with a circular root fillet. All three specimen types were tested to determine their single-

TABLE 10.2

Comparison of Directly and Traditionally Designed HCR Gears

Design method	Traditional (baseline)		Direct			
Tooth shape	Symmetric		Symmetric		Asymmetric	
Gear mesh						
Gear	Pinion	Gear	Pinion	Gear	Pinion	Gear
Number of teeth	27	49	27	49	27	49
Module, mm	3.0		3.0		3.0	
X-shift	0.15	−0.15	N/A	N/A	N/A	N/A
Root fillet profile	Full Circle		Optimized			
Operating pressure angle	20°		21.5°		24°/16°*	
Drive contact ratio	2.04		2.04		2.04/2.48*	
Pitch diameter, mm	81.0	147.0	81.0	147.0	81.0	147.0
Outer diameter, mm	89.40	153.60	89.108	154.442	89.576	154.963
Root diameter, mm	73.658	137.822	73.040	138.347	72.394	137.901
Root clearance, mm	0.371	0.389	0.259	0.273	0.322	0.262
Tooth thickness at pitch diameter, mm	5.040	4.384	5.0140	4.410	5.020	4.404
Tooth thickness at outer diameter, mm	1.18	1.70	1.12	1.13	1.01	1.02
Center distance, mm	114.0		114.0		114.0	
Face width, mm	30	30	30	30	30	30

(*Continued*)

TABLE 10.2 (Continued)

Design method	Traditional (baseline)		Direct			
Tooth shape	Symmetric		Symmetric		Asymmetric	
Gear mesh						
Gear	Pinion	Gear	Pinion	Gear	Pinion	Gear
Driving torque, Nm	300		300		300	
Bending stress, MPa	147	150	122(−17%)	123(−18%)	126(−14%)	126(−16%)
Contact stress, MPa		824	808(−2%)		774(−6%)	
Drive flank specific sliding velocity	0.367	0.323	0.342	0.342	0.336	0.336
Bearing load, N		7882	7961(+1%)		8108(+3%)	
Maximum tooth tip deflection, mm	0.0071	0.0068	0.0073	0.0070	0.0084	0.0081

Note
* Drive/coast flank

tooth bending (STBF) fatigue characteristics relative to the baseline specimens with a circular root fillet form. The baseline symmetric and asymmetric gear specimens were tested for scuffing resistance comparison. The gear test specimens are presented in Figure 10.1.

The objective of this work was to evaluate the potential benefits of asymmetric involute gear teeth and symmetric teeth with the optimized root fillet geometry for helicopter main transmission applications. It involved not only quantifying performance improvements achieved by these concepts but evaluating the practicality of manufacturing gears with asymmetric teeth and optimized root fillet geometry for aerospace applications.

(a)

(b)

(c)

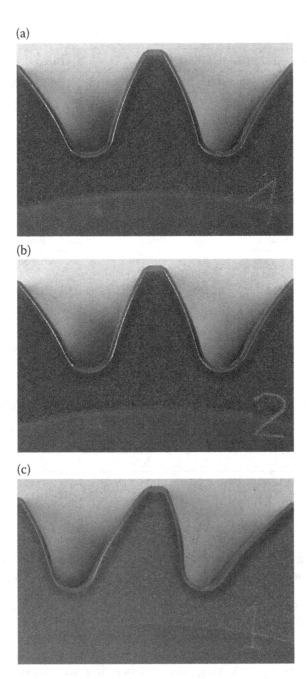

FIGURE 10.1
Test specimen gear tooth profiles; (a) baseline gear teeth, (b) symmetric gear teeth with optimized fillet, (c) asymmetric gear teeth. (Courtesy of Boeing Co, Philadelphia, Pennsylvania. With permission.)

10.2.1 Test Specimen Design and Analysis

Test specimen gears designed for this program were representative of helicopter main drive gears in diametral pitch, pressure angle, material, and processing. Standardized traditional toothed designs have been developed for bending fatigue and scuffing test rigs that Boeing Rotorcraft uses for gear research. Specimens of each type were manufactured using aerospace production techniques and requirements. A manufacturing approach was developed to reduce material and processing variability. The test specimen gear designs were analyzed to predict their bending and contact stresses and compared to stresses predicted for the baseline test specimens.

The single tooth bending fatigue test gears are 32-tooth gears with groups of 4 teeth removed per quadrant to allow for assembly into the single tooth bending fatigue (STBF) test fixture. The asymmetric gear tooth form for the STBF test specimens was nominally based on the standard STBF gear specimen. It enabled the asymmetric toothed specimen to fit the existing test fixture with only minor modifications for the tooth load angle and provides a direct comparison between asymmetric and conventional gears of the same pitch diameter and face width. The form of the optimized root fillet profile was determined analytically under the condition of having the same root diameter as that in the baseline gears. Figure 10.2 shows a comparison of the circular fillet and the optimized root fillet geometries.

Table 10.3 presents the gear parameters, and FEA calculated bending stresses for the STBF test gears.

Similarly, the scuffing test gears are within the design experience range of typical main transmission helicopter power gears. Table 10.4 presents the gear parameters for the scuffing test gear specimens.

10.2.2 Test Specimen Manufacturing

The asymmetric gear specimens, optimized root fillet gear specimens, and baseline circular fillet test gears were fabricated by Aero Gear (South Windsor, Connecticut). The specimens were fabricated from aerospace quality (AGMA Grade 3) 9310 steel with all pertinent records and certifications retained. All specimens were low pressure carburized and high pressure gas quenched. Low pressure carburizing and high pressure gas quench heat treating processes were performed at Solar Atmospheres (Souderton, Pennsylvania). The material for all specimens was from the same lot, and the heat treatment processes, grinding stock removal, and shot peening processes for all specimens were identical. All gears were surface temper etch inspected and magnetic particle inspected after the completion of machining.

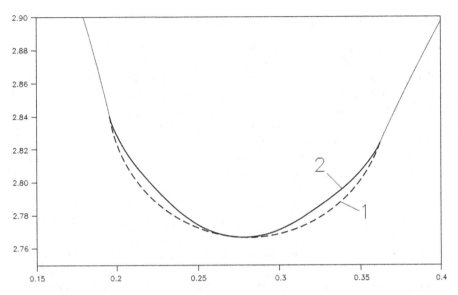

FIGURE 10.2
Coordinate plot of tooth fillet design geometries; 1: circular fillet, 2: optimized fillet. (From Brown, F.W., et al., Analysis and testing of gears with asymmetric involute tooth form and optimized fillet form for potential application in helicopter main drives. *AGMA Fall Technical Meeting*, Milwaukee, Wisconsin, October 18–19, 2010, 10FTM14. With permission.)

TABLE 10.3

STBF Test Gear Specimen Parameters

	Symmetric gears with circular fillets (baseline)	Symmetric gears with optimized fillets	Asymmetric gears with circular fillets
Number of teeth of both mating gears	32	32	32
Diametral pitch, 1/in	5.333	5.333	5.333
Pressure angle	25°	25°	35°/15°*
Pitch diameter, in	6.000	6.000	6.000
Base diameter, in	5.4378	5.4378	4.9149/ 5.7956*
Outside diameter, in	6.3975	6.3975	6.3864
Root diameter, in	5.571	5.571	5.558
Form diameter, in	5.6939	5.6939	5.6581/5.8110*
Circular tooth thickness, in	0.2895	0.2895	0.2895
Face width, in	0.375	0.375	0.375
Torque, in – lb	5000	5000	5000
Load application radius, in	3.06	3.06	3.06
Calculated maximum bending stress, psi	57887	48387 (−16.4%)	54703 (−5.5%)

Note
* Drive/coast flank

TABLE 10.4

Scuffing Test Gear Specimen Parameters

	Symmetric gears with circular fillets (baseline)	Asymmetric gears with circular fillets
Number of teeth of both mating gears	30	30
Diametral pitch, 1/in	5.000	5.000
Pressure angle	25°	35°/18°
Pitch diameter, in	6.000	6.000
Base diameter, in	5.4378	4.9149/5.7063[*]
Outside diameter, in	6.400 max	6.403 max
Root diameter, in	5.459 max	5.510
Form diameter, in	5.6864	5.6415/5.7607[*]
Circular tooth thickness, in	0.3096	0.3096
Face width, in	0.50	0.50
Drive contact ratio	1.417	1.25
Torque, in – lb	6000	6000
Calculated maximum contact stress, psi	193180	174100 (-9.9%)

Note
[*] Drive/coast flank

All specimens produced for this project were ground using conventional gear tooth form grinding equipment, including the asymmetric tooth specimens and specimens with optimized root fillet geometry. The form grinding process is often used to grind conventional symmetric gear teeth with circular fillets in helicopter main drives. The CBN (Cubic Boron Nitride) form grinding wheels were produced from data shown on the engineering drawings for both the asymmetric gear teeth and optimized fillet geometry. An example of the CBN gear grinding setup is shown in Figure 9.4. Measurements of the gear teeth including the fillet profile were carried out using conventional CMM gear checking equipment and software.

10.2.3 Test Arrangement and Procedure

Single tooth bending fatigue tests were performed on non-rotating single tooth bending fatigue test fixtures, shown in Figure 10.3. These fixtures are loaded by Baldwin-Lima Hamilton IV-20 universal fatigue machines through a series of alignment fixtures and in-line load cells. These fatigue machines are capable of 18,000 lb (10,000 lb steady load and 8000 lb alternating load).

For the STBF testing of the subject gears, pulsating fatigue load is applied to the tooth through the load link and test fixture arrangement shown in Figure 10.3. The test gear teeth were cycled at approximately 1,200 cycles per minute. Before the start of testing, alignment of the fixture was verified with a strain-gauged baseline specimen. The specimen was instrumented with three strain gauges across the face width and was used to align the fixture as well as correlate load applied to stress in the fillet of the tooth. For fatigue testing, each tested tooth is instrumented with a crack-wire, as seen in Figure 10.4. Upon failure of the crack-wire due to the presence of a fatigue crack, the test machine is triggered to shut down. The crack-wire is placed so that a crack length of 0.050 inches is detected. Magnetic particle inspection is used to confirm the presence of a crack. Each tooth specimen was run continuously until failure or runout. For this project, runout was defined as 1.0×10^7 cycles.

Scuffing tests of asymmetric gear specimens and baseline specimens were conducted on a gear research test stand. The test stand is a split-coupling torque design. The test gears are outboard of the main housing and can be quickly inspected or changed by removal of a simple cover (Figure 10.5).

FIGURE 10.3
STBF test fixture with asymmetric gear installed (Courtesy of Boeing Co, Philadelphia, Pennsylvania. From Brown, F.W., et al., Analysis and testing of gears with asymmetric involute tooth form and optimized fillet form for potential application in helicopter main drives. *AGMA Fall Technical Meeting*, Milwaukee, Wisconsin, October 18–19, 2010, 10FTM14. With permission.)

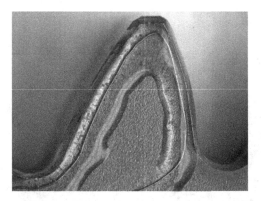

FIGURE 10.4
Asymmetric STBF test tooth with crack-wire installed (Courtesy of Boeing Co, Philadelphia, Pennsylvania. From Brown, F.W., et al., Analysis and testing of gears with asymmetric involute tooth form and optimized fillet form for potential application in helicopter main drives. *AGMA Fall Technical Meeting*, Milwaukee, Wisconsin, October 18–19, 2010, 10FTM14. With permission.)

A separate lubrication system serves the test specimen chamber, which was isolated from the test stand drive lubrication system. The lubricant supply to the test gears could be heated or cooled to supply lubricant at a constant temperature to the test gears. The test gears were subjected to a series of 15-minute, incrementally loaded runs. At the end of each 15-minute run, a visual evaluation of the test gear teeth was conducted. If the condition of the gears did not meet the criteria for scuffing failure, the next higher incremental load was applied. This procedure was continued until a scuffing failure was observed. For purposes of this test program, a scuffing failure was declared when 25% of the available tooth contact surface exhibited visible evidence of radial scratch marks, characteristic of scuffing, on a minimum of 10 teeth.

10.2.4 Test Results

After the single tooth bending fatigue tests, all crack locations were verified both visually and using magnetic particle inspection (MPI), as shown in Figure 10.6. Cracks were also opened to determine the origins and confirm the validity of the results. In Figure 10.7 the dark dashed line represents the extent of fatigue propagation, and the arrow indicates the fracture origin.

FIGURE 10.5
Scuffing test rig with cover removed and test specimen gears installed (Courtesy of Boeing Co, Philadelphia, Pennsylvania. From Brown, F.W., et al., Analysis and testing of gears with asymmetric involute tooth form and optimized fillet form for potential application in helicopter main drives. *AGMA Fall Technical Meeting*, Milwaukee, Wisconsin, October 18–19, 2010, 10FTM14. With permission.)

Fatigue results for the single tooth bending fatigue tests of the asymmetric tooth, the optimized root fillet tooth, and the baseline specimens are presented in Figure 10.8. Curves for the optimized root fillet data and the asymmetric data were assumed to be parallel to the baseline curve.

Figures 10.9 and 10.10 present the typical scuffing failures. These figures show the vertical scratches indicative of a scuffing failure, associated with the breakdown of the separating lubricant film between the gears.

Figure 10.11 shows the scuffing results for baseline and asymmetric gears. The 35° pressure angle asymmetric gears showed an improvement of approximately 25% in mean scuffing load (torque) compared to the baseline symmetric tooth specimens. The Mean – 3 Sigma levels are also shown, based on a population of 8 baseline gear data points and 6 asymmetric gear data points.

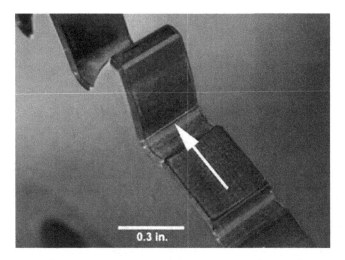

FIGURE 10.6
Cracked STBF test gear tooth – showing MPI crack indication (Courtesy of Boeing Co, Philadelphia, Pennsylvania. From Brown, F.W., et al., Analysis and testing of gears with asymmetric involute tooth form and optimized fillet form for potential application in helicopter main drives. *AGMA Fall Technical Meeting*, Milwaukee, Wisconsin, October 18–19, 2010, 10FTM14. With permission.)

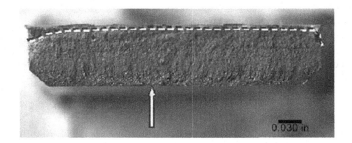

FIGURE 10.7
Fractograph of STBF test tooth (Courtesy of Boeing Co, Philadelphia, Pennsylvania. From Brown, F.W., et al., Analysis and testing of gears with asymmetric involute tooth form and optimized fillet form for potential application in helicopter main drives. *AGMA Fall Technical Meeting*, Milwaukee, Wisconsin, October 18–19, 2010, 10FTM14. With permission.)

10.2.5 Results Analysis

The STBF test results shown in Figure 10.8 indicate the asymmetric tooth gear design mean endurance limit was significantly higher, on the order of 16% higher, than the mean endurance limit of the baseline symmetric tooth design. It should be pointed out that there are relatively few data points, four failure points, and one runout (included as a failure point in

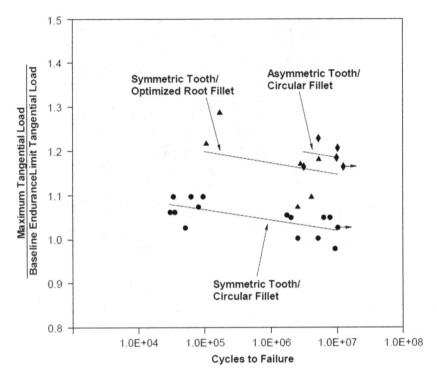

FIGURE 10.8
STBF data for asymmetric gears and optimized root fillet gears along with baseline symmetric tooth/circular fillet test data (From Brown, F.W., et al., Analysis and testing of gears with asymmetric involute tooth form and optimized fillet form for potential application in helicopter main drives. *AGMA Fall Technical Meeting*, Milwaukee, Wisconsin, October 18–19, 2010, 10FTM14. With permission.)

the data analysis) for the asymmetric tooth specimens. Nonetheless, the results of this testing indicate that asymmetric teeth offer an improvement in bending fatigue strength, although additional testing would serve to refine the magnitude of the improvement. It is interesting to note that the FE analysis of the asymmetric tooth STBF design predicted a 5.5% reduction in maximum bending stress compared to the baseline symmetric design.

The STBF results for the optimized fillet geometry design showed an improvement in mean gear tooth bending fatigue strength exceeding 10%, based on limited testing – six failure points. The data points for these tests display more variation (scatter) than either the baseline data or the asymmetric tooth data. Post-test evaluation of the test specimens and observations of the fracture surfaces did not indicate any anomalies which could explain

FIGURE 10.9
Scuffing failure of baseline test gear (Courtesy of Boeing Co, Philadelphia, Pennsylvania. From Brown, F.W., et al., Analysis and testing of gears with asymmetric involute tooth form and optimized fillet form for potential application in helicopter main drives. *AGMA Fall Technical Meeting*, Milwaukee, Wisconsin, October 18–19, 2010, 10FTM14. With permission.)

the variation, such as variations in optimized fillet form/dimensions or specimen metallurgy. One theory is that the test fixture was damaged while testing at the higher load levels. The FEA of the optimized fillet design indicated a reduction in maximum bending stress (calculated) of 16.4% compared to the baseline circular fillet design. While not tested in this project, the combination of asymmetric teeth and optimized fillet geometry together, in the same gear design, may offer improvements in tooth bending fatigue strength greater than either of the concepts taken individually. The decision was made early in this project to test each concept separately. The reasoning was that if one concept or the other proved to be impractical from a manufacturing standpoint, data of value would still be attained for the other concept. Since both concepts appear viable from a manufacturing standpoint, their combination in one gear design is worth further investigation.

The scuffing test results (Figure 10.11) indicated an improvement in mean scuffing load (torque) to failure of 25% for the asymmetric tooth gear specimens compared to the baseline symmetric tooth specimens. The improvement in calculated Mean-3 Sigma scuffing performance is even better. Although based on limited testing – 8 baseline gear points and 6 asymmetric

FIGURE 10.10
Close-up view of a representative scuffed tooth (Courtesy of Boeing Co, Philadelphia, Pennsylvania. From Brown, F.W., et al., Analysis and testing of gears with asymmetric involute tooth form and optimized fillet form for potential application in helicopter main drives. *AGMA Fall Technical Meeting*, Milwaukee, Wisconsin, October 18–19, 2010, 10FTM14. With permission.)

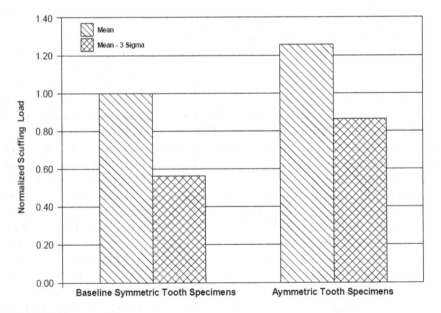

FIGURE 10.11
Results of baseline symmetric and asymmetric gear scuffing tests (From Brown, F.W., et al., Analysis and testing of gears with asymmetric involute tooth form and optimized fillet form for potential application in helicopter main drives. *AGMA Fall Technical Meeting*, Milwaukee, Wisconsin, October 18–19, 2010, 10FTM14. With permission.)

tooth gear data points – this is a very significant improvement in scuffing resistance due to asymmetric gear tooth geometry. This improvement was in the primary drive direction of the asymmetric teeth. The opposite (coast) direction scuffing performance of the asymmetric teeth was not tested in this project. This improvement in scuffing resistance can be utilized to advantage in high speed, scuffing critical gear applications.

10.2.6 Testing Results Conclusion

Test results demonstrated higher bending fatigue strength for both the asymmetric tooth form and optimized fillet form compared to baseline designs. Scuffing resistance was significantly increased for the asymmetric tooth form compared to a traditional symmetric involute tooth design.

10.3 Design Method Selection

Analytical and experimental comparison of traditional and direct approaches to gear design indicates certain benefits of Direct Gear Design® for custom gear drives. These benefits mainly include bending stress reduction that increases tooth strength, and contact stress reduction that increases the tooth surface endurance and wear resistance reducing the pitting and scuffing probability. Stress reduction provided by advanced tooth geometry allows the boosting of gear drive power transmission density, increasing its load capacity or/and reducing its size and weight, prolonging its life, and improving reliability. Possible stress and gear size reduction lead to potential cost reduction by using a reduced amount of and/or less expensive materials.

Application of an asymmetric tooth profile with a higher pressure angle, besides the stress reduction, provides lower specific sliding velocities and higher thickness of the elastohydrodynamic lubricant film (because of larger tooth contact curvature radii) on the drive flanks of the gear teeth. This leads to increased gear efficiency. Independent parameter selection of the drive and coast flanks and the root fillet profile of asymmetric teeth makes it also possible to reduce gear tooth stiffness, noise, and vibration.

Now a gear designer has a choice to make: which gear design approach is more suitable for a particular gear drive application. Table 10.5 summarizes the main characteristics of both traditional and Direct Gear Design® methods.

TABLE 10.5

Summary of Traditional and Direct Gear Design®

Traditional gear design	Direct Gear Design®
Basic principle	
Gear design is driven by standards and manufacturing convenience	Gear design is driven by application, and product technical and market performance requirements
Advantages	
• Universal applicability; • Availability of standard design manuals, software, and tooling; • Gear interchangeability; • Low tooling inventory; • Vast accumulated experience and testing database.	High gear drive performance provided by bending and contact stress reduction include: • High power transmission density – high load capacity or/and reduced size and weight; • Longer life; • Lower cost; • Lower noise and vibration; • Higher efficiency and reliability.
Drawbacks	
Limited gear drive performance.	Requires custom dedicated tooling for every gear. Limited experience and testing database. Absence of commercially available design manuals and software.
Typical applications	
Standardized gears and gear drives: • Stock gears; • Gearboxes with interchangeable gear sets; • Low production volume machined gears.	Custom gear drives: • Formed gears fabricated by plastic and metal injection molding, powder metal processing, die cast, net forging, etc.; • High production volume machined gears; • Gear drives with special requirements and extreme applications (aerospace, racing, automotive transmissions, etc.).

Further development of the Direct Gear Design® method and testing database accumulation should encourage its implementation in custom gear transmission.

11

Direct Gear Design® Applications

The Direct Gear Design approach is implemented in many custom gear drives. This chapter describes three such implementation examples presenting different applications, gear tooth geometries, materials, and fabrication technologies.

11.1 Speed Boat Gearbox

The Marine Technology Inc. turbine race boat is powered by two Lycoming T55 turboshaft engines required light and compact gearboxes. The boat power train arrangement also necessitated significant vertical offset between the turbine and propeller shafts. This offset defined the gearbox arrangement/envelope with two idler gears. Figure 11.1 shows the gearbox schematic arrangement.

11.1.1 Gear Design

Gearbox data:

- Lycoming T55 turboshaft engine maximum power – 3,000 HP;
- Turbine shaft RPM – 16,000;
- Gear ratio – 2:1;
- Vertical input/output shaft offset – 21.20″;
- Overall gearbox dimensions (length × height × width) – 16″ × 34″ × 5″;
- Gearbox weight – 270 lbs.

The gearbox was designed and manufactured by Three Sigma Manufacturing, Inc. (Kent, Washington) that subcontracted AKGears, LLC (Shoreview, Minnesota) for optimized gear design. The high contact ratio (HCR) spur gear tooth geometry was chosen to provide high load capacity, low vibration level, and zero axial thrust load on the bearings. Symmetric gear tooth profiles were selected because of two idler gears that have both tooth flanks equally loaded. The tooth root fillets were optimized to minimize bending stress concentration. Table 11.1 presents the gear design data and the stress analysis results.

Gear tooth profiles are shown in Figure 11.2.

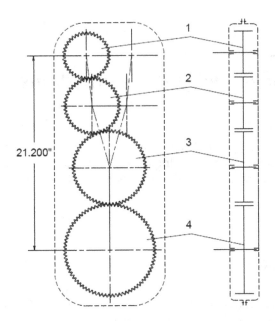

FIGURE 11.1
Gearbox arrangement; 1: input gear, 2: idler gear #1, 3: idler gear #2, 4: output gear.

11.1.2 Gear Fabrication

Considering a low production volume (only two gearboxes required), Three Sigma Manufacturing has applied its unique gear wire EDM fabrication technology. Gears were made out of the AISI 9310 steel, carburized and hardened with the surface hardness 58–62 HRC, core hardness 34–42 HRC, and final carburized depth .020–.030″. The wire cutting process, although proprietary, can be summarized as follows: the gear blanks were roughly machined, and heat treated to required core hardness. The gear teeth were preliminary wire cut leaving only 0.005″ stock. Then gears were carburized and quenched. The tooth tips were not carburized. The final tooth geometry was then cut by the wire EDM process, with special positioning techniques used to ensure uniform stock removal. A 10 µin (microinches) Ra surface finish was achieved. The internal spline is also produced in the same fixture, providing for very close concentricity to the gear teeth. The gears were then treated with the REM process [106], which improved the surface finish on the tooth flanks to 4 µin Ra.

11.1.3 Gearbox Performance Testing

Performance testing was done with the gearbox installed in the race boat. Gearbox operation was assessed by its temperature, vibration level, and

TABLE 11.1

Speedboat Gearbox Data

Gear	Input Gear	Idler Gear #1	Idler Gear #2	Output Gear
	Gear Geometry Data			
Number of Teeth	40	48	64	80
Diametral Pitch, in	8.000	8.000	8.000	8.000
Pressure Angle, °	21.0	21.0	21.0	21.0
Pitch Diameter, in	5.000	6.000	8.000	10.000
Base Diameter, in	4.6679	5.6015	7.4686	9.3358
Tooth Tip Diameter, in	5.321/ 5.326	6.326/6.331	8.315/8.323	10.307/10.315
Maximum Form Diameter, in	4.763	5.758	7.7385	9.7368
Root Diameter, in	4.623/ 4.629	5.629/5.635	7.632/7.638	9.633/9.641
Tooth Thickness at Pitch Diameter, in	0.1906/ 0.1931	0.1906/0.1931	0.1906/0.1931	0.1906/0.1931
Tooth Tip Radius, in	0.010/0.013	0.010/0.013	0.010/0.013	0.010/0.013
Tooth Tip Land, in	0.028/0.040	0.028/0.040	0.035/0.049	0.042/0.055
Face Width, in	1.650	1.650	1.550	1.550
Center Distance, in	5.5000+/−.0025	7.0000+/−.0025		9.0000+/−.0025
	Accuracy and Inspection Parameters			
Accuracy Grade per AGMA 2000-A88	Q11B	Q11B	Q11B	Q11B
Runout Tolerance, in	0.0012	0.0012	0.0013	0.0014
Pitch Variation, in	+/−0.0003	+/−0.0003	+/−0.0003	+/−0.0003
Profile Tolerance, in	0.0004	0.0004	0.0004	0.0004
Lead Tolerance, in	0.0004	0.0004	0.0004	0.0004
Pin Diameter, in	0.250	.250	.250	.250
Measurement Over Pins, in	5.406/5.412	6.408/6.414	8.411/8.416	10.412/10.418
	Tolerance Analysis Results			
Operating Pressure Angle, °	20.90/21.10	20.92/21.08		20.93/21.04
Operating Contact Ratio	2.01/2.11	2.04/2.15		2.02/2.15
Operating Normal Backlash, in	0.003/0.013	0.003/0.013		0.003/0.014
Radial Clearance, in	0.016/0.029	0.018/0.032	0.020/0.034	0.017/0.030
	Stress Analysis Results			
Power, HP	3000			
RPM	16,000	13,333	10,000	8,000
Torque, in-lb	11,657	13,988	18,651	23,314
Bearing Load, lb	4,995	4,995	4,995	4,995
Bending Stress, psi	35,920	34,380	36,720	36,530
Contact Stress, psi	142,200	133,470		115,423
Gear Material	AISI 9310 (carburized, harden)			

(a)   (b)

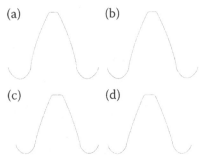

(c)   (d)

FIGURE 11.2
Tooth profiles; (a) input gear, (b) idler gear #1, (c) idler gear #2, (d) output gear.

overall vehicle performance. There were no major maintenance issues, the temperature was in the range considered safe, and vibration was minimal. The vehicle ultimately exceeded 210 mph numerous times, and the gearbox project was deemed a success. Figures 11.3–11.6 present the photos of the input gear teeth, gears, mounted gearbox, and the racing boat.

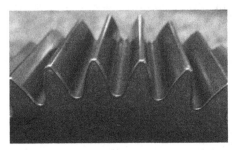

FIGURE 11.3
Input gear teeth (Courtesy of Three Sigma Manufacturing, Inc., Kent, Washington. With permission.)

FIGURE 11.4
Gears in mesh (Courtesy of Three Sigma Manufacturing, Inc., Kent, Washington. With permission.)

FIGURE 11.5
Gearbox (Courtesy of Three Sigma Manufacturing, Inc., Kent, Washington. With permission.)

FIGURE 11.6
Race boat (Courtesy of Three Sigma Manufacturing, Inc., Kent, Washington. With permission.)

11.2 Turboprop Engine Gearbox

Apparently, the first production application of asymmetric tooth gears in the aviation industry was for the TV7–117S turboprop engine gearbox [107–109]. The engine and gearbox were developed by Klimov Corporation (St. Petersburg, Russia) with the assistance of the Central Institute of Aviation Motors (CIAM, Moscow, Russia) for a commuter airplane Ilyushin Il-114 and produced by Chernyshev Enterprise (Moscow, Russia). Table 11.2 presents the main characteristics of the gearbox.

Figure 11.7 shows the TV7–117S gearbox arrangement. This arrangement was successfully used in an older generation of Russian turboprop engines

TABLE 11.2

TV7–117S Turboprop Engine Data

Input Turbine RPM	17500
Output Prop RPM	1200
Total Gear Ratio	14.6:1
Overall Dimensions, mm:	520
• Diameter	645
• Length	
Gearbox weight, N	1050
Maximum Output Power, hp	2800

AI-20 and AI-24, and it was able to provide high power transmission density for a required gear ratio. The first planetary differential stage has three planet gears. The second "star" type coaxial stage has five planet (idler) gears and a stationary planet carrier. The first stage sun gear is connected to the engine turbine shaft. Its ring gear is connected to the second stage sun gear, and its planet carrier is connected to the second stage ring gear and output propeller shaft. This arrangement makes it possible to transmit about 33% of the engine power through the first stage carrier directly to the propeller shaft, bypassing the second stage. It allows a reduction in size and weight of the second stage that transmits only about 67% of the engine power from the first stage ring gear to the second stage sun gear, and then through the planet gears to the second stage ring gear that is attached to the propeller shaft.

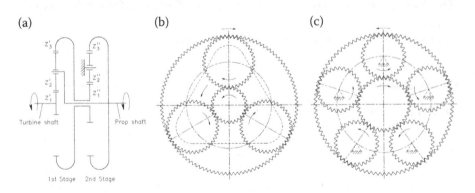

FIGURE 11.7
Gearbox arrangement (a), first (b) and second (c) stages with rotation directions (view from the input shaft). ((a) from Novikov, A.S., et al., Application of gears with asymmetric teeth in turboprop engine gearbox. *Gear Technology*, January/February 2008, 60–65. With permission.)

Table 11.3 presents the gear geometry and accuracy parameters, and operating torques and stresses. All gears were made out of forged blanks of 20KH3MVF (EI-415) steel. Its chemical composition includes: Fe – base material; C – 0.15–0.20%; S – <0.025%; P – <0.030%; Si – 0.17–0.37%; Mn – 0.25–0.50%; Cr – 2.8–3.3%; Mo – 0.35–0.55%; W – 0.30–0.50%; Co – 0.60–0.85%; and Ni – <0.5%.

The sun and planet gears were machined with the protuberance hobs before carburizing and quenching. A custom protuberance hob (Figure 11.8) was used to provide a final cutting of the gear tooth root fillet, leaving the grinding stock only on the tooth involute flanks. This allowed avoiding possible grinding "hotspots" that could propagate initial cracks in the tooth root area. Unground tooth root fillet retains residual compressive stress (about 300–600 MPa [111]) after heat treatment, which increases tooth bending strength. Table 11.4 shows the protuberance gear hob parameters.

Figure 11.9 shows the 1st stage sun gear tooth profile after the protuberance hobbing.

The ring gear involute flanks were preliminarily machined with a special asymmetric tooth shaper cutter. Then the form disk mill cutter was used to cut the root fillets. Figure 11.10 shows the 1st stage ring gear tooth space profile after machining and before carburizing and heat treatment. Table 11.5 presents its parameters.

Unlike that for the symmetric tooth gears, proper positioning of the asymmetric tooth gear blanks relative to the cutting tool is critical. Otherwise, the drive flank of one gear can be positioned in contact with the coast profile of the mating gear, which will make a gear assembly impossible. After the tooth cutting, the gears are carburized and heat treated to achieve a tooth surface hardness of >59 HRC with the case depth of 0.6–1.0 mm. The tooth core hardness is 33–45 HRC. The MAAG HSS-30 and HSS-60 generating gear grinding machines were used for final gear tooth flank grinding of the sun and planet gears. Simultaneous processing of both flanks of asymmetric teeth required a special grinding machine setup. The tip/root relief profile modification was applied to the driving tooth flanks of the sun gears and both flanks of the planet gears (Figure 11.11). Table 11.6 presents the flank modification parameters.

Assembly of the gearbox includes the selection of the planet gears and their initial angular orientation based on the runout and pitch variation. Based on these parameters, all planet gears were divided into several groups. Then during assembly position and orientation of each planet gear depended on its runout and pitch variation, providing a better load distribution between planet gears, and also reducing gear noise and vibration [8].

Application of the asymmetric teeth provided a low weight to output torque ratio and a significantly reduced noise and vibration level, and cut down duration and expense of operational development. A cross section of

TABLE 11.3

Gear Data

Stage	1st			2nd		
Gear	**Sun**	**Planet**	**Ring**	**Sun**	**Planet**	**Ring**
			Gear Geometry Data			
Number of Gears	1	3	1	1	5	1
Nominal Number of Teeth	28	41	107	38	31	97
Module, mm	3.000	3.000	3.000	3.362	3.362	3.362
Nominal Drive Pressure Angle	33°	33° and 25°	25°	33°	33° and 25°	25°
Nominal Coast Pressure Angle	25°	25° and 33°	33°	25°	25° and 33°	33°
Nominal Pitch Diameter, mm	84.000	123.000	321.000	127.756	104.222	326.114
Drive Base Diameter, mm	70.448	103.156 and 111.476	290.925	107.145	87.408 and 94.457	295.560
Coast Base Diameter, mm	76.130	111.476 and 103.156	269.213	115.786	94.457 and 87.408	273.502
Tooth Tip Diameter, mm	90.02/90.16	128.44/128.60	323.88/324.11	134.07/134.23	110.93/111.07	329.67/329.90
Root Diameter, mm	76.55/77.05	114.55/115.05	337.50/337.70	118.56/119.06	95.45/95.95	345.00/345.20
Tooth Thickness at Nominal Pitch Diameter, mm	4.773/4.814	4.325/4.365	0.667/0.621	4.972/5.018	5.253/5.299	1.104/1.059
Tooth Tip Radius, mm	0.20/0.40	0.20/0.40	0.30/0.50	0.20/0.40	0.20/0.40	0.30/0.50
Tooth Tip Land, mm	0.630/0.982	0.596/0.960	0.395/0.806	0.711/1.081	0.717/1.074	0.350/0.761
Face Width, mm	34.75/35.00	31.75/32.00	25.48/26.00	37.75/38.00	34.75/35.00	27.48/28.00
Center Distance, mm		103.50±0.01			116.00±0.01	
			Accuracy and Inspection Parameters			
Accuracy Grade per GOST 1643-81 [110]	5-5-4B	5-5-4B	5-5-4B	5-5-4B	5-5-4B	5-5-4B
Runout Tolerance, mm	0.016	0.016	0.022	0.022	0.016	0.022

Pitch Variation, mm	±0.006	±0.006	±0.007	±0.007	±0.006	±0.007
Profile Tolerance, mm	0.006	0.006	0.007	0.007	0.006	0.007
Lead Tolerance, mm	0.0055	0.0055	0.0055	0.0055	0.0055	0.0055
Pin Diameter, mm	6.000	6.000	6.000	7.000	7.000	7.000
Measurement Over Pins, mm	93.754 / 93.819	131.991 / 132.057	320.476 / 320.554	138.929 / 139.004	115.665 / 115.738	325.309 / 325.385

Tolerance Analysis Results

Operating Drive Pressure Angle	32.98°/33.02°		29.87°/29.93°	32.99°/33.03°	29.88°/29.93°	29.88°/29.93°
Operating Coast Pressure Angle	24.97°/25.03°		36.64°/36.68°	24.98°/25.04°	36.65°/36.68°	36.65°/36.68°
Operating Drive Contact Ratio	1.18/1.26		1.33/1.36	1.20/1.28	1.35/1.38	1.35/1.38
Operating Coast Contact Ratio	1.33/1.42		1.18/1.21	1.36/1.44	1.21/1.24	1.21/1.24
Operating Normal Backlash, mm	0.196/0.322		0.197/0.406	0.189/0.320		0.206/0.414
Radial Clearance, mm	0.651/ 1.033	0.936/ 1.308	0.953/ 1.376	0.921/ 1.159	0.978/ 1.356	0.948/ 1.336	0.898/ 1.321	0.936/ 1.164

Stress Analysis Results

Maximum Power, HP: 2800

RPM	17500	−11132	−3063	−3063	3755	1200
Torque per Mesh, Nm	374	548	1430	858	700	2191
Bending Stress, MPa — Tension	240	257	280	298	318	289	345	371
Bending Stress, MPa — Compression	−384	−306	−403	−355	−500	−470	−385	−430
Contact Stress, MPa	960		604	1043		809

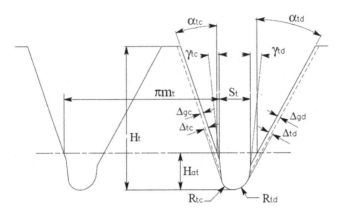

FIGURE 11.8
Protuberance gear hob profile.

TABLE 11.4

Protuberance Gear Hob Profile Data

Stage		1st		2nd	
Gear		**Sun**	**Planet**	**Sun**	**Planet**
Number of teeth		28	41	38	31
Hob module, mm	m_t	2.876	2.905	3.224	3.224
Hob tooth thickness at pitch line, mm	S_t	2.356	2.480	2.312	2.521
Hob tooth addendum, mm	H_{at}	2.124	2.157	1.858	2.124
Minimal whole depth, mm	H_t	7.80	7.80	8.20	8.20
Drive profile angle, °	α_{td}	29.0	30.0	29.0	29.0
Coast profile angle, °	α_{tc}	19.064	20.631	19.064	19.064
Drive protuberance angle, °	γ_{td}	6.0	6.0	6.0	6.0
Coast protuberance angle, °	γ_{tc}	6.0	6.0	6.0	6.0
Drive protuberance offset, mm	Δ_{td}	0.29	0.31	0.31	0.31
Coast protuberance offset, mm	Δ_{tc}	0.29	0.31	0.31	0.31
Drive grinding stock, mm	Δ_{gd}	0.17	0.20	0.20	0.20
Coast grinding stock, mm	Δ_{gc}	0.17	0.20	0.20	0.20
Drive side tip radius, mm	R_{td}	1.00	1.00	1.20	1.10
Coast side tip radius, mm	R_{tc}	0.65	0.70	0.80	0.80

the first-stage sun gear is shown in Figure 11.12. Photos of the gears and gear assemblies of the TV7–117S gearbox are shown in Figures 11.13–11.18. Figure 11.19 shows the Ilyushin Il-114 commuter airplane with the TV7–117S engines.

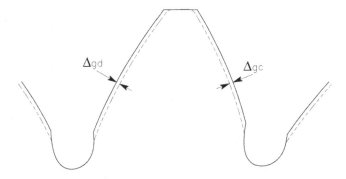

FIGURE 11.9
The 1st stage sun gear tooth profile after the protuberance hobbing; Δ_{gd} and Δ_{gc}: drive and coast flank grinding stocks.

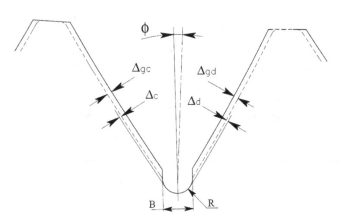

FIGURE 11.10
The 1st stage ring gear tooth space profile after machining; Δ_{gd} and Δ_{gc}: drive and coast flank grinding stocks; Δ_d and Δ_c: drive and coast flank undercuts; R: root fillet radius; B: fillet cut width; ϕ: fillet angle.

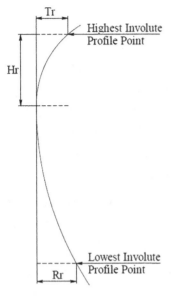

FIGURE 11.11
Tooth flack modification chart. *Hr*: tip relief height; *Tr*: tip relief depth; *Rr*: root relief depth.

TABLE 11.5

Ring Gear Root Fillet Data

Stage		1st	2nd
Gear		ring	ring
Number of teeth		107	97
Fillet radius, mm	R	0.60/0.70	0.65/0.75
Fillet cut width, mm	B	1.20/1.40	1.30/1.50
Fillet angle, °	ϕ	3.4	3.3
Drive flank fillet undercut, mm	Δ_d	0.06/0.08	0.06/0.08
Coast protuberance offset, mm	Δ_c	0.06/0.08	0.06/0.08
Drive grinding stock, mm	Δ_{gd}	0.19/0.20	0.19/0.20
Coast grinding stock, mm	Δ_{gc}	0.19/0.20	0.19/0.20

TABLE 11.6

Tooth Flank Modification Parameters

Stage		1st		2nd	
Gear		Sun	Planet	Sun	Planet
Number of teeth		28	41	38	31
Tip relief height, mm	H_r	2.5/3.5	2.0/3.0	3.0/4.0	2.0/3.0
Tip relief depth, mm	T_r	0.006/0.010	0.002/0.006	0.008/0.012	0.003/0.008
Root relief depth, mm	R_r	0.008/0.014	0.002/0.006	0.010/0.017	0.002/0.008

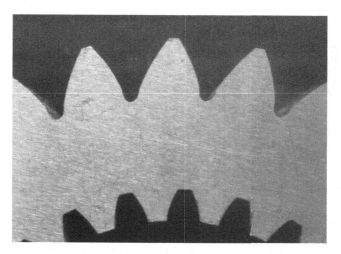

FIGURE 11.12
First stage sun gear cross section. (Courtesy of Chernyshev Enterprise, Moscow, Russia. From Novikov, A.S., et al., Application of gears with asymmetric teeth in turboprop engine gearbox. *Gear Technology*, January/February 2008, 60–65. With permission.)

FIGURE 11.13
First stage sun gear. (Courtesy of Chernyshev Enterprise, Moscow, Russia. From Novikov, A.S., et al., Application of gears with asymmetric teeth in turboprop engine gearbox. *Gear Technology*, January/February 2008, 60–65. With permission.)

FIGURE 11.14
First stage assembly. (Courtesy of Chernyshev Enterprise, Moscow, Russia. From Novikov, A.S., et al., Application of gears with asymmetric teeth in turboprop engine gearbox. *Gear Technology*, January/February 2008, 60–65. With permission.)

FIGURE 11.15
Second stage sun gear. (Courtesy of Chernyshev Enterprise, Moscow, Russia. From Novikov, A.S., et al., Application of gears with asymmetric teeth in turboprop engine gearbox. *Gear Technology*, January/February 2008, 60–65. With permission.)

FIGURE 11.16
Second stage planet gear carrier assembly. (Courtesy of Chernyshev Enterprise, Moscow, Russia. From Novikov, A.S., et al., Application of Gears with Asymmetric Teeth in Turboprop Engine Gearbox. *Gear Technology*, January/February 2008, 60–65. With permission.)

FIGURE 11.17
Second stage ring gear.

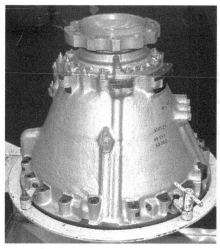

FIGURE 11.19
Ilyushin Il-114 commuter airplane.

FIGURE 11.18
Assembled gearbox. (Courtesy of Chernyshev Enterprise, Moscow, Russia. From Novikov, A.S., et al., Application of Gears with Asymmetric Teeth in Turboprop Engine Gearbox. *Gear Technology*, January/February 2008, 60–65. With permission.)

11.3 Modernization of Helicopter Gearbox

This section describes the modernization of the main gearbox for a new Ukrainian helicopter SMB-2 "Nadiya" (Figure 11.20) with asymmetric tooth gears [112]. This helicopter is a deep modification of the Russian light multipurpose Mil-2 helicopter. The gearbox modernization was required because of the replacement of the GTD-350 turboshaft engines with the new AI-450M engines, manufactured by the Motor Sich Company (Zaporizhia, Ukraine). In comparison to its predecessor, AI-450M has 465 HP power output (vs 400 HP of GTD-350), its fuel consumption is 27% lower, and it is 25 kg lighter [113].

Instead of a complete redesign of the main gearbox, the task of transmitting the increased engine power output was solved by the implementation of asymmetric tooth gears in the most loaded second and third stages of the gearbox (Figure 11.21). The power of the two turboshaft engines is transmitted by their turbine shafts 1 through the bevel gears z_1, z_2, and then the spur gears z_3, z_4, and z_5 to the bull gear z_6 connected to the main propeller shaft 2. About 25% power of one engine is used to drive the tail propeller through the spur gear z_7 and the bevel gears z_8, z_9. The original

FIGURE 11.20
MSB-2 "Nadia" helicopter.

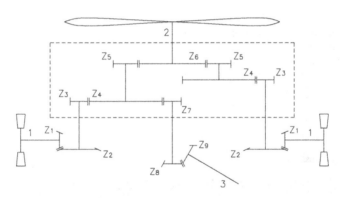

FIGURE 11.21
Main gearbox arrangement, modernized gears are shown in the dash line box; 1: engine turbine shafts, 2: main propeller shaft, 3: shaft to the tail propeller; z_1: first stage bevel pinions, z_2: first stage bevel gears, z_3: second stage pinions, z_4: second stage gears, z_5: third stage pinions, z_6: third stage gear, z_7: tail drive spur gear, z_8: tail drive bevel pinion, z_9: tail driven bevel gear. (Courtesy of Motor Sich JSC, Zaporizhia, Ukraine. From Shankin, S.I. et al., Modernization of main helicopter gearbox with asymmetric tooth gears. Gear Solutions, March 2014, 46–49. With permission.)

bevel gear pairs had sufficient load capacity to withstand the increased power of new engines and did not require any modernization. The second and third stage spur gears had to be redesigned to reduce contact stress levels and to achieve sufficient tooth flank surface durability.

The original spur symmetric tooth gears were designed traditionally using preselected basic racks and their addendum modifications (X-shifts). The modernized spur gears with asymmetric teeth were constructed utilizing the Direct Gear Design® method [31].

Tables 11.7 and 11.8 present the main geometric parameters of the original and modernized spur gears.

TABLE 11.7

Geometric Parameters of Original and Modernized Gears of Second Stage

Gear Geometry	Original		Modernized	
Design Method	Rack Generation		Direct Gear Design®	
Tooth Profile	Symmetric		Asymmetric	
Gear	Pinion	Gear	Pinion	Gear
Number of Teeth	15	59	15	59
Center Distance	206.504		206.504	
Basic Rack Module	5.500	5.500	N/A	N/A
Basic Rack Pressure Angle	20°	20°	N/A	N/A
Addendum Modification	0.5	0.0751	N/A	N/A
Operating Module	5.581	5.581	5.581	5.581
Operating Drive Pressure Angle	22.177°		32°	
Operating Coast Pressure Angle	22.177°		20°	
Drive Contact Ratio	1.43		1.30	
Coast Contact Ratio	1.43		1.59	

TABLE 11.8

Geometric Parameters of Original and Modernized Gears of Third Stage

Gear Geometry	Original		Modernized	
Design Method	Rack Generation		Direct Gear Design®	
Tooth Profile	Symmetric		Asymmetric	
Gear	Pinion	Gear	Pinion	Gear
Number of Teeth	14	37	14	37
Center Distance	206.504		206.504	
Basic Rack Module	8.000	8.000	N/A	N/A
Basic Rack Pressure Angle	28°	28°	N/A	N/A
Addendum Modification	0.5	0.0751	N/A	N/A
Operating Module	8.098	8.098	8.098	8.098
Operating Drive Pressure Angle	29.28°		32°	
Operating Coast Pressure Angle	29.28°		20°	
Drive Contact Ratio	1.20		1.30	
Coast Contact Ratio	1.20		1.57	

Parameters of asymmetric teeth were selected to satisfy the following requirements:

- Drive flank pressure angle should not exceed 32° to limit the maximum radial load and keep the original bearings;
- Minimum drive flank contact ratio is 1.3;
- Minimum tooth tip thickness is 2.3 mm for the second stage gears and 3.1 mm for the third stage gears to avoid the harden through tooth tip;
- Minimum coast flank pressure angle in the second stage is 20°, because the coast flanks of one of the gears are used to transmit torque to the tail propeller;
- The tooth root fillet is not ground after heat treatment; its profile should not interfere with the grinding wheel trajectory.

Before heat treatment, the gear teeth were machined, leaving the grinding stock about 0.20 mm on the tooth flanks and providing the final root fillet profile. Unground tooth root fillet retains residual compressive stress after heat treatment, which increases tooth bending strength. This required gear hobs with a protuberance. However, due to the low number of teeth of the second and third stage pinions, the protuberance hobs could not provide the required tooth root fillet surface finish. To solve this issue, the pinion flanks were machined with conventional hobs (without a protuberance), and then, before the gear heat treatment, the pinion root fillets were ground. Figure 11.22 shows the second stage pinion tooth profiles before and after heat treatment.

Before heat treatment, the driven gears of both the second and third stages were machined with the protuberance hobs. Figure 11.23 shows the third stage gear tooth profiles before and after heat treatment.

Tables 11.9 and 11.10 show the stress analysis results.

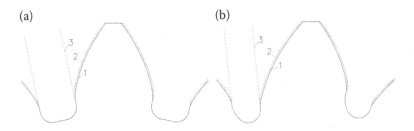

FIGURE 11.22
Second stage pinion tooth profiles; (a) original symmetric, (b) asymmetric; 1: after heat treatment and flank grinding, 2: before treatment, 3: root fillet grinding wheel. (Courtesy of Motor Sich JSC, Zaporizhia, Ukraine. From Shankin, S.I. et al., Modernization of main helicopter gearbox with asymmetric tooth gears. *Gear Solutions*, March 2014, 46–49. With permission.)

(a)   (b)

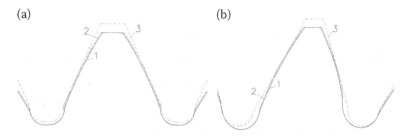

FIGURE 11.23
Third stage gear tooth profiles; a: original symmetric, b: asymmetric; 1: after heat treatment and flank grinding, 2: before treatment, 3: profile of hob with protuberance. (Courtesy of Motor Sich JSC, Zaporizhia, Ukraine. From Shankin, S.I. et al., Modernization of main helicopter gearbox with asymmetric tooth gears. *Gear Solutions*, March 2014, 46–49. With permission.)

TABLE 11.9
Stress Analysis Results of Original and Modernized Gears of Second Stage

Gear Geometry	Original		Modernized	
Design Method	Rack Generation		Direct Gear Design®	
Tooth Profile	Symmetric		Asymmetric	
Gear	Pinion	Gear	Pinion	Gear
Number of Teeth	15	59	15	59
Face Width	58.0	55.0	58.0	55.0
Torque, Nm	1,255	4,936	1,255	4,936
Max. Contact Stress, MPa	1308		1186	
Contact Stress Reduction	–		9%	
Max. Root Bending Stress, MPa	305	298	279	287
Root Bending Stress Reduction	–	–	9%	4%

All gears are made of forged blanks of 16H3NVFMB (VKS-5, DI-39) steel. Its chemical composition includes: Fe – base material, C – 0.13–0.19%, Si – 0.17–0.37%, Mn – 0.50–0.90%, Cr – 2.65–3.25%, Ni – 0.4–0.8%, S – <0.025%, P – <0.025%, Cu – < 0.30%.

The gear blanks' position during machining must provide the asymmetric teeth pointed in a definite direction (clockwise or counterclockwise). Otherwise, the drive flank of one gear will be positioned in contact with the coast profile of the mating gear, which would make gearbox assembly impossible.

TABLE 11.10

Stress Analysis Results of Original and Modernized Gears of Third Stage

Gear Geometry	Original		Modernized	
Design Method	Rack Generation		Direct Gear Design®	
Tooth Profile	Symmetric		Asymmetric	
Gear	Pinion	Gear	Pinion	Gear
Number of Teeth	14	37	14	37
Face Width	92.0	92.0	92.0	92.0
Torque, Nm	4,936	13,046	4,936	13,046
Max. Contact Stress, MPa	1513		1425	
Contact Stress Reduction	–		6%	
Max. Root Bending Stress, MPa	367	371	352	364
Root Bending Stress Reduction	–	–	4%	2%

(a)   (b)

FIGURE 11.24
Compound gear; (a) original design with symmetric teeth, (b) modernized design with asymmetric teeth. (Courtesy of Motor Sich JSC, Zaporizhia, Ukraine. From Shankin, S.I. et al., Modernization of main helicopter gearbox with asymmetric tooth gears. *Gear Solutions*, March 2014, 46–49. With permission.)

After the tooth cutting, the gears are carburized and heat-treated to achieve a tooth surface hardness of 59–60 HRC with a case depth of 1.2–1.4 mm. The core tooth hardness is 34–42 HRC.

Final gear machining includes tooth grinding and honing. Asymmetric gears require a special setup for both these operations.

Figure 11.24 shows the original and modernized compound gears that contain the second stage gear and third stage pinion. In these compound gears, the second stage gear is driven, and the third stage pinion is driving.

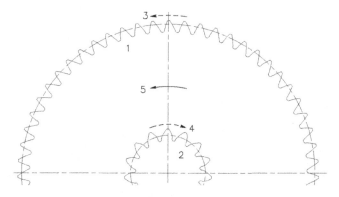

FIGURE 11.25
Asymmetric tooth slant directions of the compound gear; 1: driven second stage gear, 2: driving third stage pinion, 3: asymmetric tooth slant direction of the driven second stage gear, 4: asymmetric tooth slant direction of the driving third stage pinion, 5: compound gear rotation direction.

For the asymmetric compound gear (Figure 11.24b), this means that asymmetric tooth incline directions of the gear and pinion must be the opposite. Figure 11.25 illustrates asymmetric tooth slant directions of the compound gear. The driven second stage gear asymmetric tooth slant direction coincides with the compound gear rotation direction. The driving third stage pinion asymmetric tooth slant direction is opposite to the compound gear rotation direction. It is a critical matter for proper gearbox assembly and operation.

Application of the asymmetric tooth gears in the most loaded second and third spur gear stages of the SMB-2 "Nadiya" helicopter main gearbox provided significant performance improvement. It allowed achieving acceptable drive flank durability without an increase of major gear dimensions and utilizing more powerful and fuel-efficient AI-450M turboshaft engines without a complete redesigning of the helicopter gearbox.

Epilogue

The 2nd edition of this book includes basic principles as well as the latest developments of the Direct Gear Design® approach. Some less relevant parts have been removed or changed significantly. Practically every chapter of the book has been revised and updated with new study results.

Over the last several decades, the main contributors to gear drive improvements have been achievements in gear materials, manufacturing technology and tooling, thermal treatment, tooth surface engineering and coatings, tribology and lubricants, testing technology and diagnostics, simulation software, and gear tooth microgeometry – which defines deviation from the nominal involute surface to achieve optimal tooth contact localization for higher load capacity and lower transmission error. At the same time, the standardized shape of the modern gear tooth and its definition method, based on tooling rack generation, is a status quo long-settled in minds of gear engineers.

The main subject of this book is the maximization of gear drive performance by customization, including complete optimization, of the gear tooth geometry. It describes the Direct Gear Design® development method of nonstandard involute gears with symmetric and asymmetric teeth. This 2nd edition covers all aspects of Direct Gear Design®, from a brief historical overview to examples of custom applications of this gear design method.

As an author, I would like to express my deep and sincere gratitude to readers of this book and encourage them to utilize its content for the practical application of Direct Gear Design®.

Alexander L. Kapelevich

References

1. Dudley, D. W. 1966. *The Evolution of the Gear Art*. Alexandria, VA: AGMA.
2. Litvin, F. L. 1998. *Development of Gear Technology and Theory of Gearing*. Washington, DC: NASA.
3. McVittie, D. 1993. The European rack shift coefficient "X" for Americans. *Gear Technology*, 10(4), July/August: 34–36.
4. Goldfarb, V. I., and A. A. Tkachev. 2005. New approach to computerized design of spur and helical gears. *Gear Technology*, 22(1), January/February: 26–32.
5. Maitra G. N., 1994. *Handbook of Gear Design*, 2nd ed. New York: McGraw-Hill.
6. Jelaska, D. T. 2012. *Gears and Gear Drives*. New York: John Wiley & Sons.
7. Linke, H., J. Börner, and R. Heß. 2016. *Cylindrical Gears*. Munich: Carl Hanser Verlag GmbH & Co. KG.
8. Vulgakov, E. B. (ed.). 1981. *Aviation Gearboxes. Handbook*. Moscow: Mashinostroenie.
9. Ketov, Ch. F. 1934. *Involute Gearing*. Leningrad-Moscow: State Scientific and Technical Publishing House for Machine Building and Metalworking.
10. Lopukhov, N. P. 1941. Geometry of involute gearing. *Theory of Machine and Mechanisms*. Moscow: Moscow Aviation Institute: 144–180.
11. Vulgakov, E. B. 1974. *Gears with Improved Characteristics*. Moscow: Mashinostroenie.
12. Vulgakov, E. B., and L. M. Vasina. 1978. *Involute Gears in Generalized Parameters*. Moscow: Mashinostroenie.
13. Vulgakov, E. B. 1995. *Theory of Involute Gears*. Moscow: Mashinostroenie.
14. Kleiss, R. E., A. L. Kapelevich, and N. J. Kleiss Jr. 2001. New opportunities with molded gears. Paper Presented at the Annual Technical Meeting of the AGMA, Detroit.
15. Da Vinci, L. ~1500. Analysis of Gear Teeth – Asymmetrical Teeth (Peg Teeth), Madrid Ms. I (BNM), fol. 5r.
16. Willis, R. 1841. *Principles of Mechanism*. London: John W. Parker.
17. Reuleaux, F. 1894. *The Constructor. A Hand-Book of Machine Design*. Philadelphia, PA: H.H. Suplee.
18. Spooner, H. S. 1908. *Machine Drawing and Design for Beginners*. London: Longmans, Green.
19. Logue, C. H. 1910. *American Machinist Gear Book*. New York: McGraw-Hill Book Company.
20. Königl, K. 1912. Osterreichische patentschrift No. 62002.
21. Leutwiler, O. A., 1917. *Elements of Machine Design*. London: McGraw-Hill Book Company.
22. Kapelevich, A. L. 2018. *Asymmetric Gearing*, Boca Raton, FL: CRC Press.
23. DiFrancesco, G., and S. Marini. 1997. Structural analysis of teeth with asymmetrical profiles. *Gear Technology*, 14(5), July/August: 16–22.

24. Sfantos, G. K., V. Spitas, and T. Costopoulos. 2003. *Optimized Asymmetric Gear Tooth Design*. Athens: National Technical University of Athens (NTUA). Technical Report No. TR-SM-0307: 1–22.

25. Deng, G., T. Nakanishi, and K. Inoue. 2003. Bending load capacity enhancement using an asymmetric tooth profile. *JSME International Journal. Series C 46*, 3: 1171–1177.

26. Vulgakov, E. B., and G. V. Rivkin. 1976. Design of gears with asymmetric tooth profile. *Mashinovedenie* 5: 35–39.

27. Kapelevich, A. L. 1984. *Research and Development of Geometry of Modernized Involute Gears*. PhD Dissertation, Moscow State Technical University.

28. Kapelevich, A. L. 1987. Synthesis of asymmetric involute gearing gears. *Mashinovedenie* 1: 62–67.

29. Kapelevich, A. L. 2016. Pitch factor analysis for symmetric and asymmetric tooth gears. *Gear Solutions*, 14(12), December: 27–33.

30. Kapelevich, A. L. 2000. Geometry and design of involute spur gears with asymmetric teeth. *Mechanism and Machine Theory* 35: 117–130.

31. Kapelevich, A. L., and R. E. Kleiss. 2002. Direct Gear Design for spur and helical gears. *Gear Technology*, 19(5), September/October: 29–35.

32. Kapelevich, A. L., and E. Taye. 2012. Self-locking gears: design and potential applications. *Gear Solutions*, 10(5), May: 53–58.

33. Vulgakov, E. B., and A. L. Kapelevich. 1980. Area of existence of helical gearing. *Vestnik Mashinostroeniya* 7: 9–11.

34. Kapelevich, A. L. 2013. *Direct Gear Design*, Boca Raton, FL: CRC Press.

35. Townsend, D. P. 1967. *Dudley's Gear Handbook*. 2nd ed. New York: McGraw Hill.

36. Kapelevich, A. L., and Y. V. Shekhtman. 2010. Area of existence of involute gears. *Gear Technology*, 27(1), January/February: 64–69.

37. Alipiev, O. L., and S. D. Antonov. 2010. Asymmetric involute – lantern meshing formed by identical spur gears with a small number of teeth. *VDI-Berichte* 2018: 925–939.

38. NASA/SP-8100. 1974. *Liquid Rocket Engine Turbopump Gears*. Washington, DC: NASA.

39. Henry, Z. S. 1995. *Bell Helicopter Advanced Rotorcraft Transmission (ART) Program*. Contract Report ARL-CR-238. Washington, DC: NASA.

40. Anderson, N. E., and S. H. Loewenthal. 1984. *Efficiency of Nonstandard and High Contact Ratio Involute Spur Gears*. AVSCOM Technical Report 84-C-9. Washington, DC: NASA.

41. Yoerkie, C. A., and A. G. Chory. 1984. Acoustic vibration characteristics of high contact ratio planetary gears. *Journal of American Helicopter Society* 40: 19–32.

42. Howe, D. C., C. V. Sundt, and A. H. McKibbon. 1988. *Advanced Counter-Rotating Gearbox*. NASA Contractor Report 180883. Washington, DC: NASA.

43. Kapelevich, A. L. 2012. Asymmetric gears: parameter selection approach. *Gear Technology*, 29(4), June/July: 48–51.

44. Standard ISO 6336. 2006. *Calculation of Load Capacity of Spur and Helical Gears*. Switzerland: ISO Copyright Office.

45. Joachim, F. J., J. Börner, and N. Kurz. 2012. How to minimize power losses in transmissions, axles and steering systems. *Gear Technology*, 29(6), September: 58–66.

46. Kapelevich, A. L., and Y. V. Shekhtman. 2017. Analysis and optimization of contact ratio of asymmetric gears, *Gear Technology*, 34(2), March/April: 67–71.

47. Flodin, A., and L. Forden, 2004. Root and contact stress calculations in surface densified PM gears. *Proceedings from World PM2004 Conference*, 2: 395–400.
48. Kapelevich, A. L., and A. Flodin, 2017. Contact ratio optimization of powder metal gears, *Gear Solutions*, 15(8), August: 39–44.
49. Krueger, D. 2017. Geometry, pitting and tooth root capacity of asymmetric gears. Proceedings of the International Conference on Gears, Garching near Munich, Germany, September, 13–15.
50. Inoue, K., and T. Masuyama. 2006. Possibilities of fatigue strength simulation in reliability design of carburized gears. Paper presented at the International Conference "Power Transmissions' 06", Novi Sad, Serbia.
51. Pedersen, N. L. 2010. Improving bending stress in spur gears using asymmetric gears and shape optimization. *Mechanism and Machine Theory* 45: 1707–1720.
52. Roth, Z., and J. Opferkuch. 2017. High load carrying tooth root fillet based on a Bézier curve. Proceedings of the International Conference on Gears, Garching near Munich, Germany, September, pp. 13–15.
53. Der Hovanesian, J. et al. 1989. Gear root stress optimization using photoelastic optimization techniques. *SAE Transactions* 97 (4): 748–755. ISSN 0096-736X.
54. Pulley, F. T. et al. 2000. Method for producing and controlling a fillet on a gear. US Patent #6164880.
55. Spitas, V., T. Costopoulos, and C. Spitas. 2005. Increasing the strength of standard involute gear teeth with novel circular root fillet design. *American Journal of Applied Sciences* 2 (6): 1058–1064.
56. Sanders, A. A., D. R. Houser, A. Kahraman, J. Harianto, and S. Shon. 2011. An experimental investigation of the effect of tooth asymmetry and tooth root shape on root stresses and single tooth bending fatigue life of gear teeth. Paper presented at the 11th International ASME Power Transmission and Gearing Conference, Washington, DC, USA.
57. Roth, M. A., Z. Roth, and K. Paetzold. 2013. Developing a bionic gear root fillet contour. Proceedings of the International Conference on Gears, Garching near Munich, Germany, Vol. 2, October 7–9, pp. 1315–26.
58. Florian, L., and Roth, Z. 2019. Bionic tooth root: fatigue testing and potential on gear units. Proceedings of the International Conference on Gears, Garching near Munich, Germany, September, pp. 18–20.
59. Kapelevich, A. L., and Y. V. Shekhtman. 2003. Direct gear design: bending stress minimization. *Gear Technology*, 20(5), September/October: 44–49.
60. Kapelevich, A. L., and Y. V. Shekhtman. 2009. Tooth fillet profile optimization for gears with symmetric and asymmetric teeth. *Gear Technology*, 26(7), September/October: 73–79.
61. Rastrigin, L. A. (ed.). 1969. *Random Search Theory and Application*. Riga: Zinatne.
62. Kapelevich, A. L., and Y. V. Shekhman. 2018. Root fillet optimization of thin rim planet gears with asymmetric teeth. *Gear Solutions*, 16(3), March: 52–55.
63. Kapelevich, A. L., and Y. V. Shekhman. 2020. Optimization of asymmetric tooth root generated with protuberance hob. *Gear Solutions*, 17(6), June: 32–37.
64. Litvin, F. L., Q. Lian, and A. L. Kapelevich. 2000. Asymmetric modified gear drives: reduction of noise, localization of contact, simulation of meshing and stress analysis. *Computer Methods in Applied Mechanics and Engineering* 188: 363–390.

65. Houser, D. R. 2006. Gear mesh misalignment. *Gear Solutions*, 4(6), June: 34–43.
66. Ellen Bergseth, E., and S. Björklund. 2010. Logarithmical crowning for spur gears. *Strojniški vestnik – Journal of Mechanical Engineering* 56(4): 239–244.
67. Kapelevich, A. L., and Y. V. Shekhtman. 2016. Rating of asymmetric tooth gears. *Power Transmission Engineering*, 10(3), April: 40–45.
68. Kapelevich, A. L., and Y. V. Shekhman. 2017. Rating of helical asymmetric tooth gears. *Gear Technology*, 34(8), November/December: 78–81.
69. Standard ANSI/AGMA 2001-D04. 2004. *Fundamental Rating Factors and Calculation Methods for Involute Spur and Helical Gear Teeth*. Alexandria, VA: AGMA.
70. Adams, C. E. 1986. *Plastic Gearing*. New York: Marcel Dekker, Inc.
71. Starzhinsky, V. E., E. V. Shalobaev, et al. 1998. *Plastic Gears in Instrument Mechanisms*. Saint-Petersburg, Gomel: MPRI NASB.
72. Standard ANSI/AGMA 1006-A97. 1997. *Tooth Proportions for Plastic Gears*. Alexandria, VA: AGMA.
73. Standard ANSI/AGMA 1106-A97. 1997. *Tooth Proportions for Plastic Gears* (Metric Addition). Alexandria, VA: AGMA.
74. Standard ANSI/AGMA 920-A01. 2001. *Materials for Plastic Gears*. Alexandria, VA: AGMA.
75. VDI-Standard: VDI 2736. parts 1–4.
76. Mao, K., W. Liebig, C. J. Hooke, and D. Walton. 2010. Polymer gear surface thermal wear and its performance prediction. *Tribology International* 43 (1–2): 433–439. ISSN 0301-679X.
77. Kapelevich, A. L. 2020. Optimal polymer gear design: metal-to-plastic conversion, *Gear Technology*, 37(3), May: 40–45.
78. Flodin. A. 2018. Powder metal through the process steps. *Gear Technology*, 35(5), September/October: 50–57.
79. S. Radzevich. 2019. *Advances in Gear Design and Manufacture*. Boca Raton, FL: CRC Press, pp. 329–362.
80. Hoganas AB handbooks. https://www.hoganas.com/handbooks.
81. ANSI/ASME Y14.7. 1971. *Gear Drawing Standard – Part 1: For Spur, Spur, Helical, Double Helical and Rack*. New York: ASME.
82. ANSI/ASME Y14.7. 1971. *Gear Drawing Standard – Part 2: For Bevel and Hypoid Gears*. New York: ASME.
83. American Standard ANSI/AGMA 2015-1-A01. 2002. *Accuracy Classification System – Tangential Measurements for Cylindrical Gears*. Alexandria, VA: AGMA.
84. American Standard ANSI/AGMA 2015-2-A06. 2006. *Accuracy Classification System – Radial Measurements for Cylindrical Gears*. Alexandria, VA: AGMA.
85. International Standard ISO 1328-1. 1995. *Cylindrical Gears – ISO System of Accuracy – Part 1: Definitions and Allowable Values of Deviations Relevant to Corresponding Flanks of Gear Teeth*. Geneva: ISO.
86. International Standard ISO 1328-2. 1997. *Cylindrical Gears – ISO System of Accuracy – Part 2: Definitions and Allowable Values of Deviations Relevant to Radial Composite Deviations and Runout Information*. Geneva: ISO.
87. Zou, Z., and, E. Morse. 2001. Statistical tolerance analysis using gap space. This paper was presented at the 7th CIRP International Seminar on Computer Aided Tolerancing. ENS de Cachan, France.

88. Creveling, C. M. 1996. *Tolerance Design: A Handbook for Developing Optimal Specifications*. Upper Saddle River, NJ: Prentice Hall, p. 422.
89. Cox, N. D. 1986. *How to Perform Statistical Tolerance Analysis*. Milwaukee, WI: American Society for Quality Control.
90. Saari, O. E. 1954. How to calculate exact wheel profiles for form grinding helical gear teeth. *American Machinist*, 13: 172–175.
91. Kapelevich, A. L., A. I. Tolchenov, and A. I. Eidinov. 1985. Application of the fly cutters in experimental production. *Aviatsionnaya Promyshlennost* 3: 39–40.
92. Andersson, M., and Flodin, A. 2017. Redesigning and prototyping a six speed manual automotive transmission of powder metal gears. Proceedings of MPT2017-Kyoto, the JSME International Conference on Motion and Power Transmissions, February 28–March 3, Kyoto, Japan.
93. Flodin, A. 2016. Powder metal gear technology: a review of the state of the art. *Power Transmission Engineering*, 10(2), March: 38–43.
94. Flodin, A., and Hirsch, M. 2017. Wear investigation of finish rolled powder metal gears. Presented at APMA 2017, The 4th International Conference on Powder Metallurgy in Asia, April 09–11, Hsinchu, Taiwan.
95. Kleiss, R. E. 2006. How to achieve a successful molded gear transmission. *Gear Technology*, 23(4), July/August: 42–47.
96. Walsh, S. F. 1993. Shrinkage and warpage prediction for injection molded components. *Journal of Reinforced Plastics and Composites* 12(7): 769–777.
97. Autodesk® Moldflow® Plastic Injection Molding Simulation Software. http://usa.autodesk.com/moldflow/.
98. Kapelevich, A. L., Y. V. Shekhtman, and T. M. McNamara. 2005. Turning an art into science. *Motion System Design*, 47(8), August: 26–31.
99. McGuinn, J. 2020. Additive manufacturing – an update. *Gear Technology*, 37(4), June: 24–29.
100. Kapelevich, A. L. 1986. Measurement over pins of the gears with asymmetric teeth. *Mashinovedenie* 6: 109–110.
101. Kapelevich, A. L. 2011. Measurement of directly designed gears with symmetric and asymmetric teeth. *Gear Technology*, 28(1), January/February: 60–65.
102. Nezhurin, I. P. 1961. Calculation of the measurement over pins of the helical gears with odd number of teeth. *Vestnik Mashinostroeniya* 2:14–17.
103. Standard ANSI/AGMA 2000-A88. 1988. *Gear Classification and Inspection Handbook*. Alexandria, VA: AGMA.
104. Coy, J. J., D. P. Townsend, and E. V. Zaretsky. 1985. *Gearing*. NASA Reference Publication 1152, AVSCOM Technical Report 84-C-15. Washington, DC: NASA.
105. Brown, F. W., S. R. Davidson, D. B. Hanes, D. J. Weires, and A. L. Kapelevich. 2011. Analysis and testing of gears with asymmetric involute tooth form and optimized fillet form for potential application in helicopter main drives. *Gear Technology*, 28(4), June/July: 46–55.
106. Winkelmann, L. W., M. D. Michaud, G. Sroka, A. A. Swiglo, and D. Mahan. 2001. Chemically accelerated vibratory finishing for the virtual elimination of wear and pitting for steel gears. 01FTM7. AGMA Fall Technical Meeting, Detroit, MI.
107. Sarkisov, A. A., and E. B. Vulgakov. 2000. The new gearbox of Klimov Corporation. *Aerospace Courier* 2: 32–33.

108. Vulgakov, E. B., and A. L. Kapelevich. 2000. The gearbox for the TV7-117S turbo-prop engine. *Vestnik Mashinostroeniya* 11: 13–17.
109. Novikov, A. S., A. G. Paikin, V. L. Dorofeyev, V. M. Ananiev, and A. L. Kapelevich. 2008. Application of gears with asymmetric teeth in turboprop engine gearbox. *Gear Technology*, 25(1), January/February: 60–65.
110. GOST 1643–81. 1981. *Tolerances for Cylindrical Gears. Interstate Standard.* Moscow: IPK.
111. Yeliseev, Y. S., V. V. Krymov, I. P. Nezhurin, V. S. Novikov, and N. M. Ryzhov. 2001. *Fabrication of Gas Turbine Engine Gears.* Moscow: Vysshaya Shkola.
112. Shankin, S. I., A. A. Stupakov, I. L. Glikson, V. M. Ananiev, and A. L. Kapelevich. 2014. Modernization of main helicopter gearbox with asymmetric tooth gears, *Gear Solutions*, 12(3), March: 46–49.
113. Mi-2MSB – Ukrainian Combat Variant of the Mi-2 Helicopter. http://www. defence24.com/265849,mi-2msb-ukrainian-combat-variant-of-the-mi-2-helicopter.

Index

Page numbers followed by *f* indicate figures; those followed by *t* indicate tables.